84 PRACTICAL IC PROJECTS YOU CAN BUILD

Other TAB books by the author:

No. 812 *Radio Control For Models*
No. 790 *Twenty One Simple Transistor Radios You Can Build–From Crystal Sets to Superhets*
No. 965 *Modern Transistor Radios (Make and Use Series)*
No. 964 *Modern Crystal Radios (Make and Use Series)*
No. 958 *A Beginner's Guide to Making Electronic Gadgets*
No. 1023 *A Beginner's Guide to Designing & Building Transistor*
No. 1113 *Understanding Electronics*

No. 1142
$8.95

84 PRACTICAL IC PROJECTS YOU CAN BUILD

BY R. H. WARRING

TAB BOOKS
BLUE RIDGE SUMMIT, PA. 17214

FIRST PRINTING—AUGUST 1979

Printed in the United States
of America

Integrated Circuits How To Make Them was originally published in 1979.
Printed by permission of Lutterworth Press.

Library of Congress Cataloging in Publication Data

Warring, Ronald Horace.
84 practical IC projects you can build

Includes
1. Integrated circuits—Amateurs' manuals. I. Title.
TK9965.W383 621.381'73 79-17480
ISBN 0-8306-1142-8 pbk.

CONTENTS

LIST OF WORKING CIRCUITS

PREFACE

INTEGRATED CIRCUITS (or ICs) are the building blocks from which modern electronic circuits are assembled. They save a lot of time in construction and give better performance than similar circuits built from separate components and, above all, are incredibly space saving. In these respects they are a big step ahead of single transistors and have made it easier for amateur constructors—as well as professionals—to build working circuits.

There are thousands of different types of ICs, each of which may be adaptable to many different working circuits (although some of the more complex ones are designed with a limited range of application). This can be quite bewildering, especially knowing how and where to start. However, from the point of view of using ICs and putting them to work, there is no need at all to know the actual circuits they contain—merely what *type* of circuit they contain and how their leads or pins are connected to other components to complete a working circuit.

This is what this book is about. It explains and 'classifies' integrated circuits in simple terms. It covers the various ways in which the simplest ICs—Op-Amps—can be worked; and describes a whole range of working circuits based on selected—and inexpensive—integrated circuits. The book contains a total of 84 different working circuits.

In fact it is really a basic—and essentially practical—'course' on understanding and using integrated circuits.

Chapter One

INTRODUCTION TO INTEGRATED CIRCUITS

THE TRANSISTOR FIRST APPEARED as a working device in 1947, since which time it has been manufactured in hundreds of millions. It took a little time to realize that the same techniques used for producing individual transistors could be applied to complete circuits and sub-circuits containing both active components (e.g. diodes and transistors) and passive components (e.g. resistors and capacitors), with all necessary interconnections in a single unit familiarly known as a 'chip'.

Apart from the obvious advantage of being able to produce complete circuits and sub-circuits in this way, the cost of producing a complex circuit by photo-etching techniques is little more than that of producing individual transistors, and the bulk of the circuit can be reduced substantially since transistors in integrated circuits do not need encapsulation or canning, and resistors and capacitors do not need 'bodies'. Another advantage is the potentially greater reliability offered by integrated circuits, since all components are fabricated simultaneously and there are no soldered joints. Also performance can be improved as more complex circuits can be used where advantageous at little or no extra cost.

The next big step in integrated circuit construction was the development of microelectronic technology or extreme miniaturization of such components and integrated circuits. Photo-etching is readily suited to this, the main problem being in checking individual components for faults due to imperfections in the manufacturing process, and achieving a high yield of fault-free chips per 'wafer' manufactured. Rejection rates are liable to rise with increasing complexity of the circuit, but modern processes now achieve a very high yield.

Basically an integrated circuit consists of a single chip of silicon, typically about 1.25 mm square (0.050 inches square) in size. Each chip may contain 50 or more separate components, all interconnected (although they may contain very many less for simpler circuits). The actual manufacturing process is concerned with producing wafers, each of which may contain several hundred chips. These wafers are processed in batches, so one single batch production may be capable of producing several thousand integrated circuit chips simultaneously, involving a total of tens of thousands of components.

It is this high production yield which is responsible for the relatively low price of integrated circuits—usually substantially less than the cost of the equivalent individual components in a chip produced separately, and in the case of some chips even less than that of a single transistor. The final selling price, however, is largely governed by demand. The integrated circuit is a mass-production item and the greater the demand for a particular chip, the lower the price at which it can be sold economically.

Fig. 1.1 shows a typical—and fairly simple—IC produced as

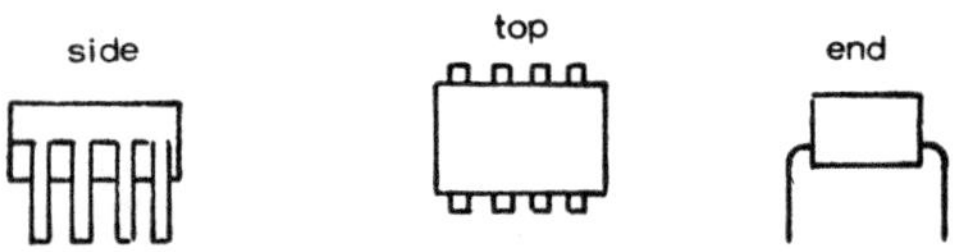

1.1 Outline shape of a typical 8-pin dual-in-line integrated circuit, about $1\frac{1}{2}$ times actual size.

a flat 'package' encapsulated in plastic. The drawing is approximately $1\frac{1}{2}$ times actual size (9.4mm long by 6mm wide). Fig. 1.2 shows the complete circuit contained in this IC, comprising 16 transistors, 8 diodes and 13 resistors. Fig. 1.3 shows the physical appearance of the chip, much magnified, when it is part of the wafer. The actual size of this chip is approximately 2.5mm by 2mm.

The actual *component density* or number of components per unit area, may vary considerably in integrated circuits. The figure of 50 components per chip has already been mentioned, which is typical of small-scale integration (SSI). It is possible to achieve much higher component densities. With medium-scale integration (MSI), component density is greater than 100 components per chip; and with large-scale integration (LSI), component density may be as high as 1000 or more components per chip. Both MSI and LSI are extensions of the original integrated circuit techniques using similar manufacturing methods. The only difference is in the matter of size and

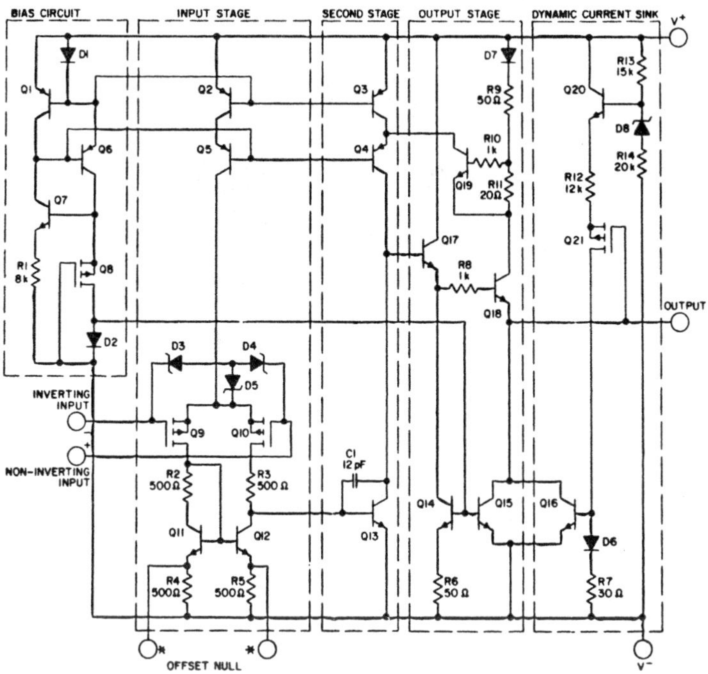

1.2 Schematic diagram of one half of a CA3240 BiMOS operational amplifier showing components and interconnections all formed in the substrate of the chip.

physical separation of the individual components and the method of inter-connection.

Monolithic and Hybrid ICs

Integrated circuits where all the components and connections are formed in the substrate of the 'chip' are known as *monolithic ICs*. There is a further class of ICs where the individual components (transistors, diodes, resistors, etc.), or even complete sub-circuits, are all attached to the same substrate but with interconnections formed by wire bonding. These are known as *hybrid ICs*. In hybrid circuits, electrical isolation is provided by the physical separation of the components.

IC Components

Transistors and *diodes* are formed directly on the surface of the chip with their size and geometry governing their electrical

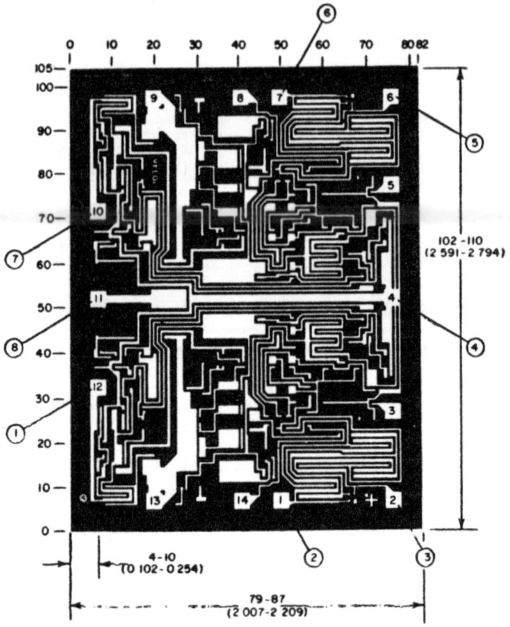

1.3 Much enlarged illustration of the CA3240 chip containing two complete circuits like Fig. 1.2. Actual dimensions of this chip are 2.5 by 2 millimetres. Grid dimensions marked around the outside of the diagram are in thousandths of an inch.

characteristics as well as density level, etc. Where a number of such components are involved in a complete integrated circuit their performance is usually better than that of a circuit with discrete (separate) components because they are located close together and their electrical characteristics are closely matched.

Resistors can be formed by silicon resistance stripes etched in the slice, or by using the bulk resistivity of one of the diffused areas. There are limits, however, to both the range and tolerance of resistance values which can be produced by these methods. 'Stripe' resistors are limited to a minimum width of about 0.025 mm (0.001 in.) to achieve a tolerance of 10 per cent. Practical values obtained from diffused resistors range from about 10 ohms to 30k ohms, depending on the method of diffusion with tolerances of plus or minus 10 per cent. Better performance can be achieved with thin-film resistors with resistance values ranging from 20 ohms to 50k ohms.

A method of getting round this problem when a high

resistance is required is to use a transistor biased almost to cut-off instead of a resistor in an integrated circuit where a resistance value of more than 50k ohms is required. This is quite economic in the case of integrated circuit manufacture and a method widely used in practice.

Capacitors present more of a problem. Small values of capacitance can be produced by suitable geometric spacing between circuit elements and utilizing the stray capacitance generated between these elements. Where rather higher capacitance values are required, individual capacitors may be formed by a reversed-bias PN junction; or as thin-film 'plate' type using a tiny aluminium plate and a MOS (metal-oxide-semiconductor) second plate. The former method produces a *polarized* capacitor and the thin film method a *non-polarized* capacitor. The main limitation in either case is the relatively low limit to size and capacitance values which can be achieved —typically 0.2pF per 0.025mm (0.001in.) square for a junction capacitor and up to twice this figure with a thin film MOS capacitor, both with fairly wide tolerances (plus or minus 20 per cent). Where anything more than moderate capacitor values are needed in an integrated circuit it is usually the practice to omit the capacitor from the circuit and connect a suitable discrete component externally.

Both resistors and capacitors fabricated in ICs also suffer from high temperature coefficients (i.e. working values varying with temperature) and may also be sensitive to voltage variations in the circuit.

Unlike printed circuits, it is not possible to fabricate inductors or transformers in integrated circuits at the present state-of-the-art. As far as possible, therefore, ICs are designed without the need for such components; or where this is not possible, a separate conventional component is connected externally to the integrated circuit.

From the above it will be appreciated that integrated circuits are quite commonly used as 'building blocks' in a complete circuit, connected to other conventional components. A simple example is shown in Fig. 1.4 using a ZN414 as a basic 'building block' in the construction of a miniature AM radio. Although a high gain device (typical power gain 72dB) the integrated circuit needs a following stage of transistor amplification to

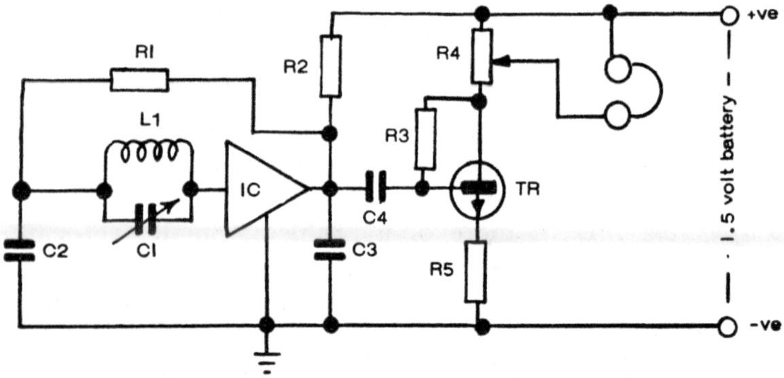

1.4 Complete radio circuit using a ZN414 integrated circuit connected to external components.

Resistors
R1 – 100 kohms
R2 – 1 kohm
R3 – 100 kohms
R4 – 10 kohms
R5 – 100 kohms

Capacitors
C1 – tuning capacitor to match tuning coil L1
C2 – 0.01 μF
C3 – 0.1 μF
C4 – 0.1 μF

L1 proprietary medium wave aerial coil on ferrite rod; or 80 turns of 30 s.w.g. enamelled copper wire wound on ferrite rod (matching value of C1 = 250 pF).

IC – ZN414
TR – ZTX300 (or equivalent)
Speaker – transistor radio crystal earpiece.

power a crystal earpiece; high value decoupling capacitors; and a standard coil and tuning capacitor for the tuned circuit. The complete circuit is capable of providing an output of 500 millivolts across the earpiece, with a supply voltage of 1.3 and typical current drain of 0.3 milliamps.

1.5 Examples of integrated circuit outlines.

A 16-pin dual in-line
B 14-pin dual in-line
C flat (ceramic) package
D 3-lead transistor 'can' shape
E 6-lead 'transistor' shape
F 8-lead 'transistor' shape
G 12-pin quad in-line with heatsink tabs

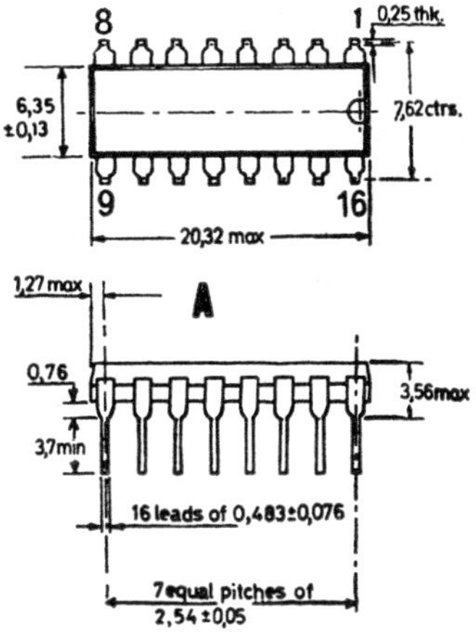

8
1
0,25 thk.
6,35
±0,13
7,62 ctrs.
9
16
20,32 max
1,27 max
A
0,76
3,56 max
3,7 min
16 leads of 0,483±0,076
7 equal pitches of
2,54 ±0,05

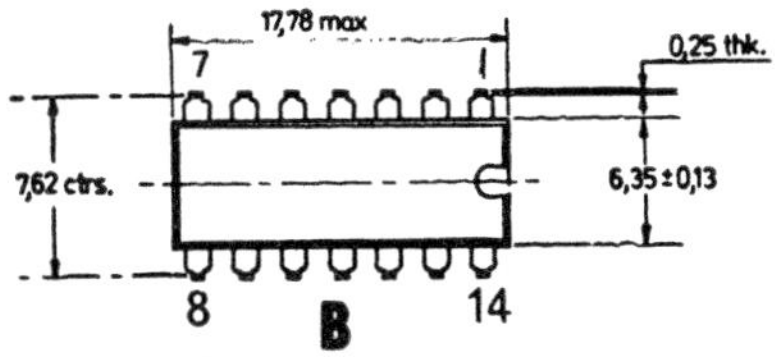

17,78 max
7
1
0,25 thk.
7,62 ctrs.
6,35±0,13
8
B
14

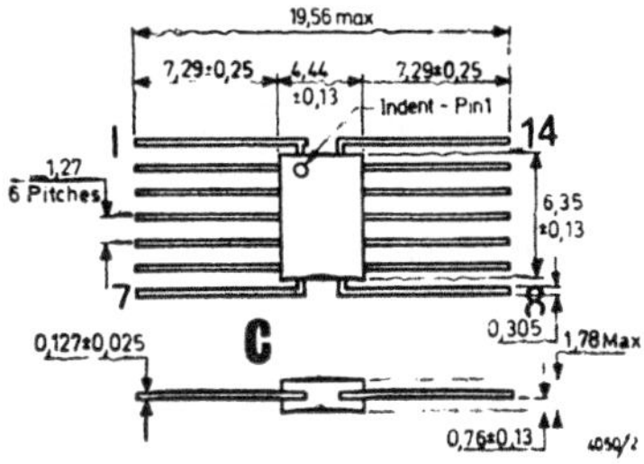

19,56 max
7,29±0,25
4,44
±0,13
7,29±0,25
Indent - Pin1
1
14
1,27
6 Pitches
6,35
±0,13
7
8
0,305
C
0,127±0,025
1,78 Max
0,76±0,13

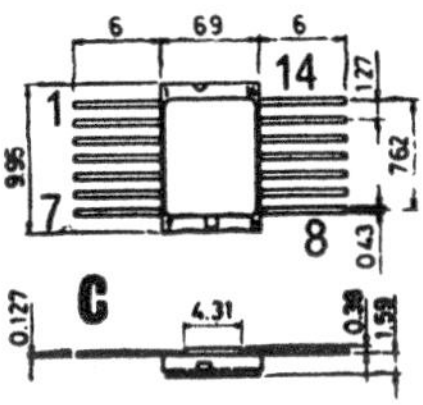

6
6.9
6
14
1.27
1
9.96
7.62
7
8
0.43
C
0.127
4.31

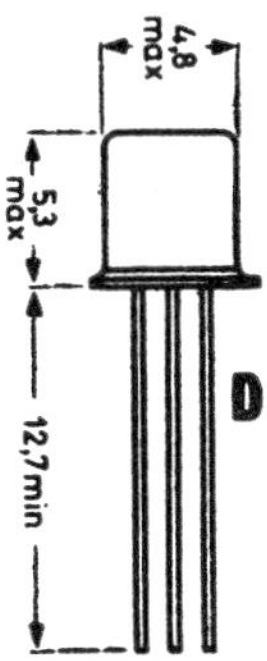

4,8
max
5,3
max
D
12,7 min

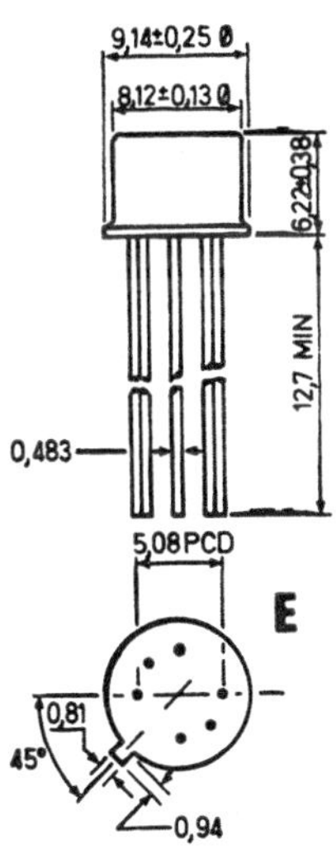

9,14±0,25 Ø
8,12±0,13 Ø
6,22±0,38
12,7 MIN
0,483
5,08 PCD
E
0,81
45°
0,94

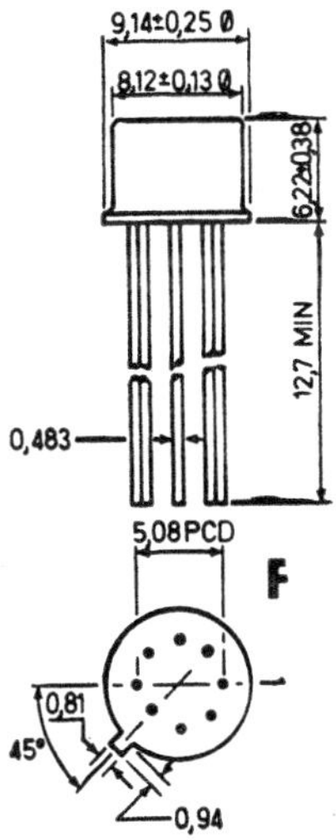

9,14±0,25 Ø
8,12±0,13 Ø
6,22±0,38
12,7 MIN
0,483
5,08 PCD
F
0,81
45°
0,94

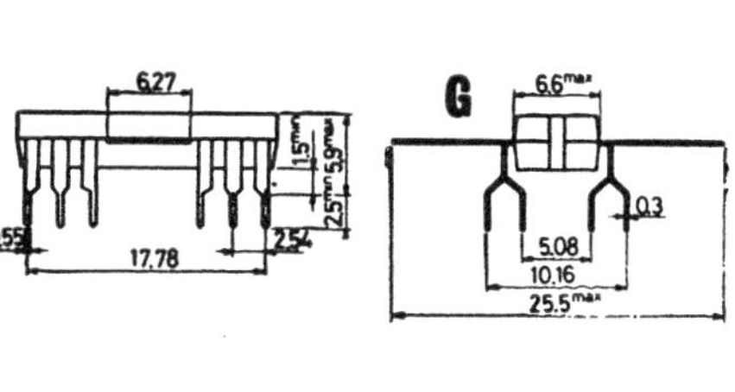

6.27
0.55
17.78
2.54
G
6.6max
0.3
5.08
10.16
25.5max

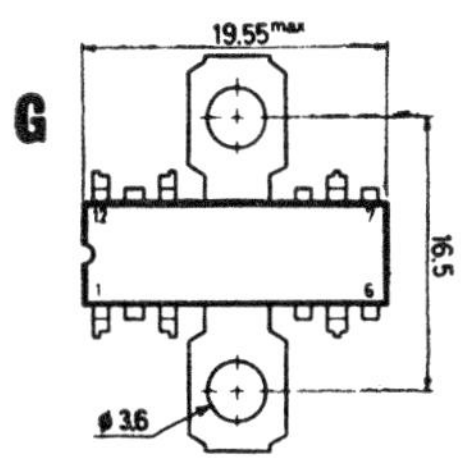

G
19.55max
12
7
1
6
16.5
ø 3.6

The Shape of ICs

ICs come in various 'package' shapes. Quite a number have the same shape (and size) as a typical transistor and are only readily identified as an IC because of the greater number of leads emerging from the bottom (a transistor usually has only three leads). These shapes are defined by the standard coding adopted for transistor outlines, e.g. TO-5, TO-18, etc., which also identifies the individual pins by numbers (e.g. *see* Fig. 1.5).

Other ICs come in the form of flat packages with leads emerging from each side. These are three different arrangements used (*see also* Fig. 1.5).

1. *Dual in-line,* where the leads on each side are bent down to form two separate rows to plug directly into a printed circuit panel or IC holder.
2. *Quad in-line,* like dual in-line, except that the leads on each side form two parallel rows.
3. *Flat,* where the leads emerge straight and from each side of the package.

In all cases leading numbering normally runs around the package, starting from top left and ending at top right (again *see* Fig. 1.5). The number of leads may be anything from eight to sixteen or even more.

Some types of holders designed to match standard pin configurations or flat shape ICs are shown in Fig. 1.6. These holders have similar pin configurations to the ICs they take. Their principal advantage is that they can be soldered to a printed circuit or Veroboard, etc., with no risk of heat damage to the IC itself since this is only plugged in after soldered connections are completed. Most circuit constructors, however, prefer to solder ICs directly to a printed circuit panel (or Veroboard).

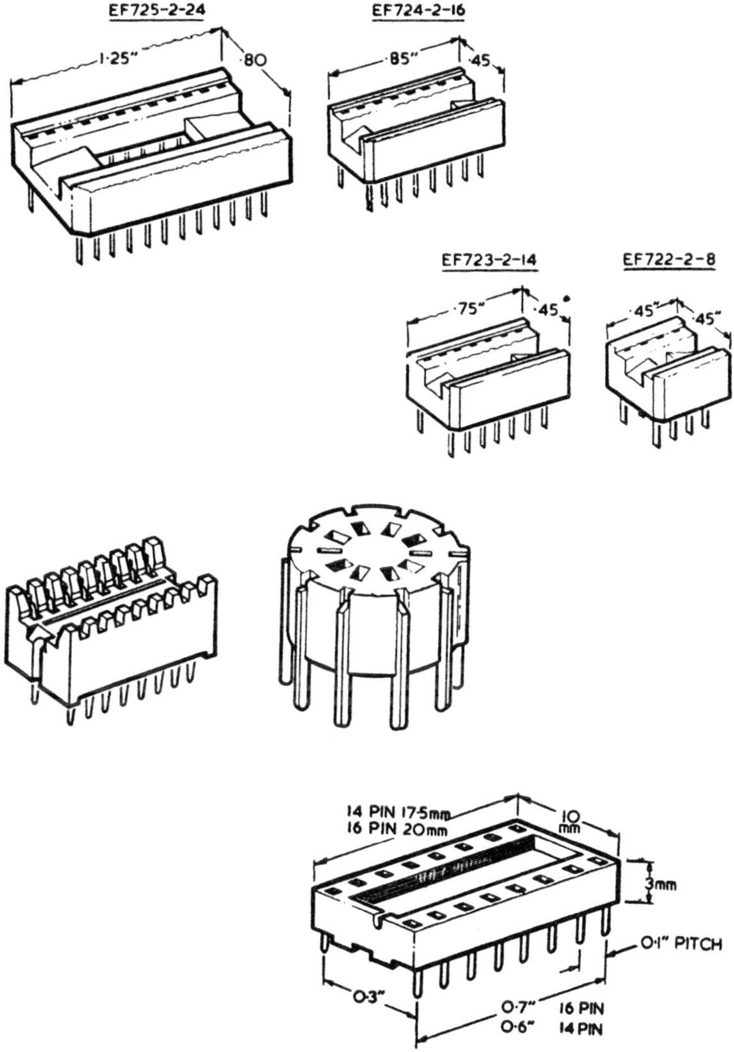

1.6 Examples of integrated circuit holders (Electrovalue).

Chapter Two

'GENERAL PURPOSE' ICs (ARRAYS)

THE DESCRIPTION 'GENERAL PURPOSE' is not accepted terminology but it is used here to describe integrated circuit chips which comprise a number of individual components, usually transistors and possibly also diodes, each component in the chip connecting to individual outlet leads. Thus by connecting to the appropriate three (or two) leads and individual transistor (or diode) it can be connected to an external circuit. Other chips of this type may also include components connected within the chip, e.g. transistors in Darlington pairs, but the same principle of application applies. The technical description of such a chip is an *integrated circuit array*.

A simple example of such a chip is shown in Fig. 2.1. It consists of three transistors (two interconnected); two types of diodes; and a Zener diode. This particular chip is used in the voltage regulator circuit described in Chapter 8 (Fig. 8.4), using two of the transistors, the SCR diode and the Zener diode.

This circuit design is shown in Fig. 2.2. The components to be utilized which are contained in the IC are enclosed in the dashed outline, i.e. TR1, TR2, D2 and D3. The other components in the chip (D1 and TR3) are not required. Resistors R1, R2, R3 and R4 and a capacitor C, are all discrete components connected externally.

Fig. 2.3 re-examines the component disposition in the chip, together with the necessary external connections. Note that the arrangement of the leads or *pin-out* arrangement does not necessarily follow the schematic diagram (Fig. 2.1) where the pins are in random order to clarify connections to internal components. The actual pin-out arrangement on ICs follows a logical order reading around the chip. Schematic diagrams may or may not follow in the same order (usually not).

Connections for completing the circuit of Fig. 2.3 are:

Leads 1, 2 and 3 are ignored as D1 is not used.

Lead 4 connects one side of the Zener diode to the common

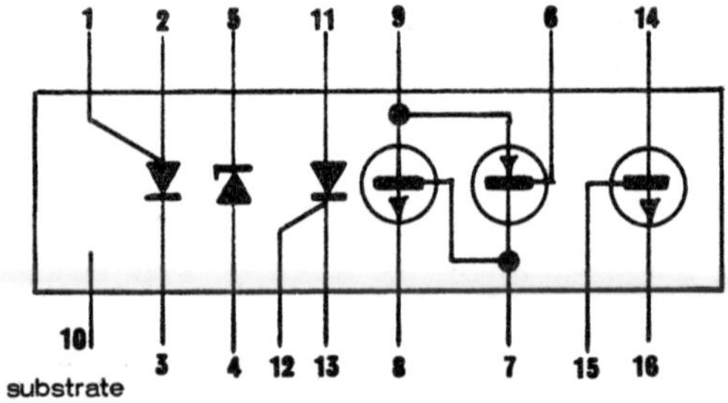

2.1 Schematic diagram of CA3097E integrated circuit array which contains two diodes, one zener diode, two NPN transistors and one PNP transistor. Numbered pin connections are also shown, these providing access to individual components in the chip. These pins are not in the physical order as found on the chip (*see* Fig. 2.3).

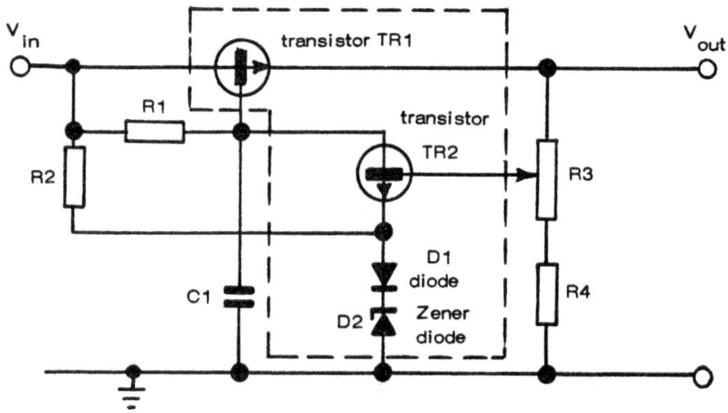

2.2 Voltage regulator circuit components within the dashed outline are in the CA3097E integrated circuit. R1, R2, R3, R4 and C are external components.

earth line and Lead 5 to Lead 13, connecting the Zener diode to the correct side of the SCR (diode).

Leads 11 and 12 connect together (as the SCR is worked as a simple diode in this circuit and the gate connection is not required).

Now to pick up the transistor connections. The base of TR1

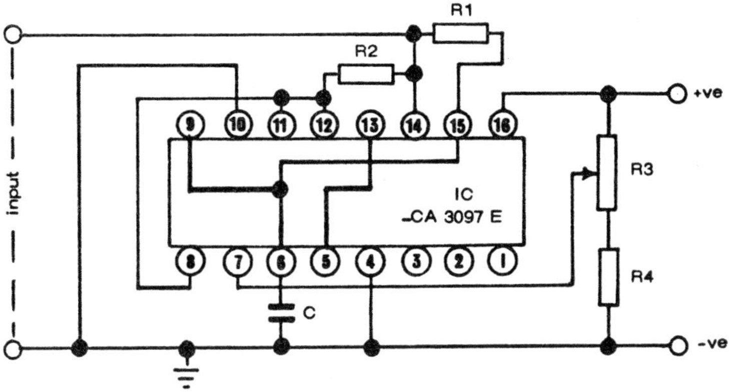

2.3 Completed voltage regulator circuit showing wiring connections made to the integrated circuit. Pins in this diagram are shown in the actual physical order they appear on the integrated circuit. For ease of reading, pins are shown numbered and enclosed in circles rather than numbered tags. On circuit drawings pin numbers may be shown circled or not.
Note. For clarity the integrated circuit is drawn much larger in proportion to the external components.

(15) connects to the external resistor R1; and the collector lead (14) to the other side of R1, which is also the input point for the circuit. The emitter lead (16) connects to output.

TR2 and TR3 in the chip are interconnected, but only one of these transistors is required. Connecting lead (6) to (9) shorts out TR2, which is not wanted. Connecting the emitter lead (8) of TR3 to 11-12 (already joined); the collector lead (9) to (6); and the base lead (7) to the centre tag of the external potentiometer R3 connects TR3 into the circuit.

It only remains for the external component connections to be completed. These are:

R2 to lead (14) and lead (12).

Capacitor C to lead (6) and earth point. Lead (10) on the IC is also the substrate or earthing point of the IC, so should also be connected to the common earth line.

One end of the potentiometer R3 to the top (output) line.

The other end of the potentiometer R3 to R4.

The other end of R4 to the bottom common earth line.

Spare Components

A number of components in an array may not be used in a particular circuit, but the cost of the single IC can often be less than that of the equivalent transistors or diodes ordered separately and used individually to complete the same circuit. The circuit using the IC will also be more compact and generally easier to construct.

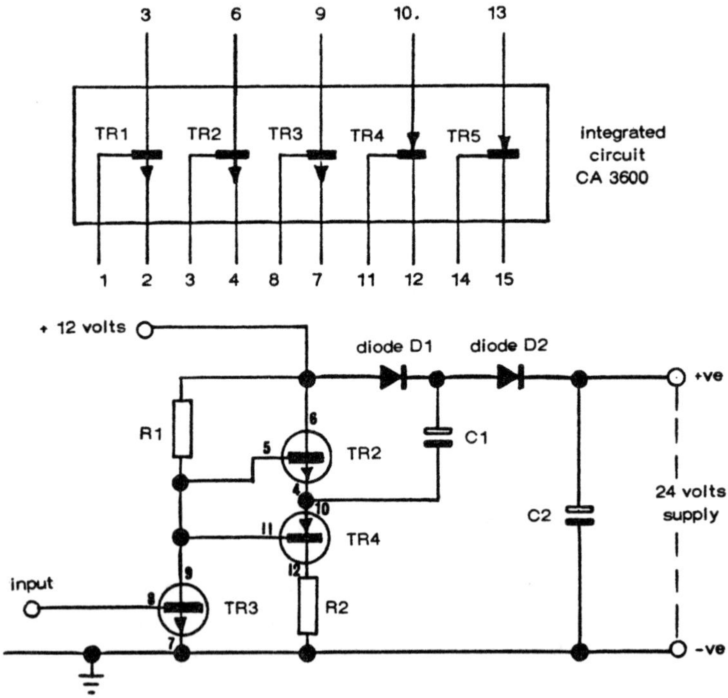

2.4 Schematic diagram of CA3600 array (*top*) and voltage doubler circuit using TR2, TR3 and TR4 from the array together with external components:

R1=10kohms
R2= 1kohm
C1=2.5μF
C2=2.5μF
diodes D1 and D2

The spare transistors in the array (TR1 and TR5) can be used instead of separate diodes, connected for diode working by ignoring the collector leads.

A little study sometimes shows where further savings are possible. Fig. 2.4, for example, shows a voltage doubler circuit for a 1 kHz square wave input signal, based on a CA3096E IC array which contains 5 transistors. Only three of these transistors are used in this particular circuit. leaving two 'spare'.

The circuit calls for two diodes D1 and D2 (as well as three resistors and two capacitors) to be added as discrete components. Transistors can also be worked as diodes (by neglecting the collector lead), and so the functions of D1 and D2 could be performed by the two 'spare' transistors in the array (thus using up all its components).

Alternatively, since the current needs a square wave input signal, the two 'spare' transistors could be used in a multivibrator circuit to provide this input, and in this case using discrete components for D1 and D2. Since diodes are cheaper than transistors, this is a more economic way of using all the components in the original array.

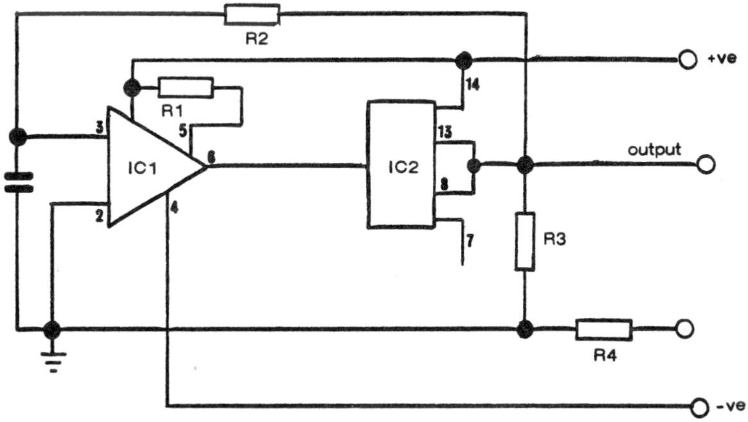

2.5 Astable multivibrator circuit using one third of CA3600E array.

IC1 — OTA CA3080
IC2 — CA3600E
R1 — 100 kohms
R2 — 5 kohms
R3 — 10 kohms
R4 — 10 kohms
C — 0.01 μF

The fact that popular ICs are quite cheap means that it is seldom worth while going to elaborate methods of trying to use all the components available in an array, unless such utilization is fairly obvious, as above. Using only part of an array can still show savings over the purchase of individual components for many circuits.

The *astable multivibrator* circuit shown in Fig. 2.5, for example, only uses one of the three complete switching circuits contained in the CA3600E array, associated with an OTA CA3080 integrated circuit and four external resistors. On the other hand, Fig. 2.6 shows a high gain amplifier circuit using all the components in the CA3600E array with external resistors.

Constant Current Circuit

A useful circuit employing the CA3018 integrated circuit array is shown in Fig. 2.7. This array comprises four transistors (two interconnected as a super-alpha pair) and four diodes.

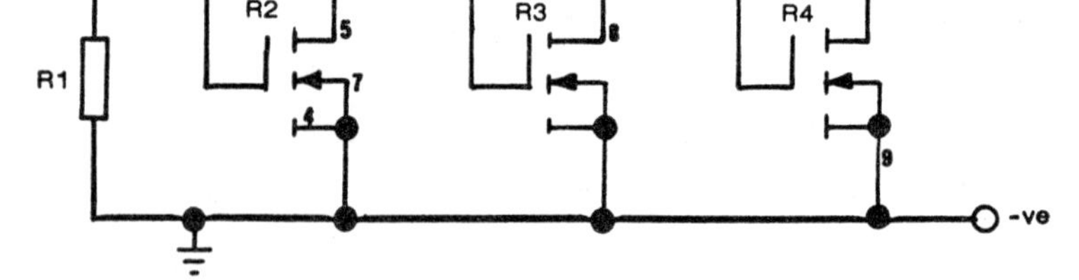

2.6 High-gain amplifier circuit using the complete CA3600E integrated circuit together with external components:

R1 – 1 Mohm
R2 – 22 Mohms
R3 – 22 Mohms
C1 – 1 μF
C2 – 1 μF
C3 – 1 μF

Tapping the super-alpha pair of transistors, a *constant current* source can be produced, the magnitude of this current being set by adjustment of the potentiometer R1 over a range of about 0.2 mA to 14 mA, depending on the actual supply voltage.

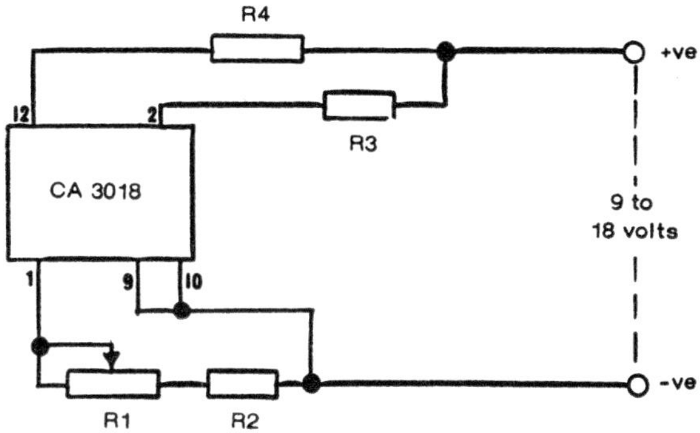

2.7 Constant current circuit using components found in CA3018 array.
External component values:

R1 – 10 kohms potentiometer
R2 – 470 ohms
R3 – 3.3 kohms
R4 is the resistance of the load through which the constant current is to flow.

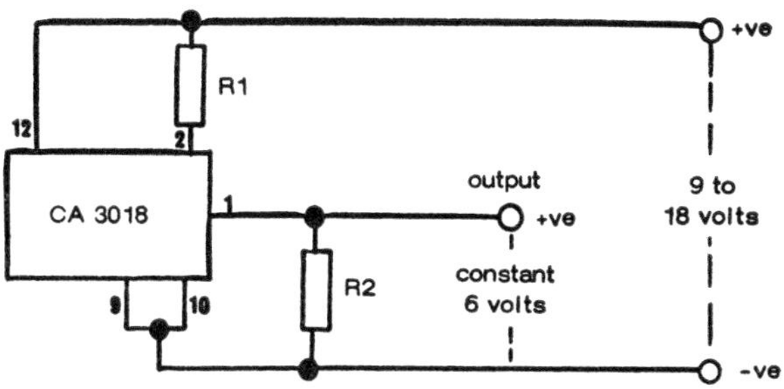

2.8 Circuit giving a constant 6 volts output from a 9 to 18 volts supply voltage, again using the CA3018 array.
External Component values:

R1 – 3.3 kohms
R2 – 4.7 kohms

The same integrated circuit can also be used as a *constant voltage* source—Fig. 2.8. In this case the constant voltage output is the Zener voltage of the transistor worked as a Zener diode, which is approximately 6 volts.

Chapter Three

OP-AMPS

OP-AMPS (OPERATIONAL AMPLIFIERS) are a particular class of integrated circuit comprising a directly-coupled high-gain amplifier with overall response characteristics controlled by feedback. The op-amp gets its name from the fact that it can be made to perform numerous mathematical operations. An op-amp is the basic building block in analogue systems and is also known as a *linear integrated circuit* because of its response.

It has an extremely high gain (theoretically approaching infinity), the actual value of which can be set by the feedback. Equally the introduction of capacitors or inductors in the feedback network can give gain varying with frequency and thus determine the operating condition of the whole integrated circuit.

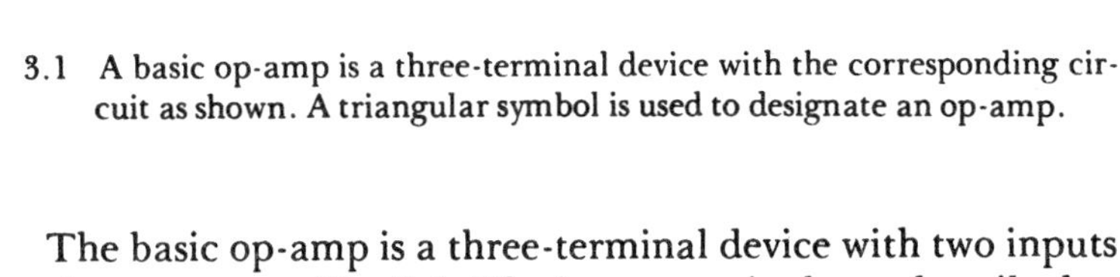

3.1 A basic op-amp is a three-terminal device with the corresponding circuit as shown. A triangular symbol is used to designate an op-amp.

The basic op-amp is a three-terminal device with two inputs and one output—Fig. 3.1. The input terminals are described as 'inverting' and 'non-inverting'. At the input there is a virtual 'short circuit', although the feedback keeps the voltage across these points at zero so that no current flows across the 'short'. The simple circuit equivalent is also shown in Fig. 3.1, when the voltage gain is given by a ratio of the impedances Z2/Z1.

Some examples of the versatility of the op-amp are given in the following simple circuits:

Adder (Fig. 3.2)

Input signals V_1, V_2 .. V_n are applied to the op-amp as shown through resistors R1, R2 . . R_n. The output signal is then a combination of these signals, i.e. giving the *sum* of the inputs.

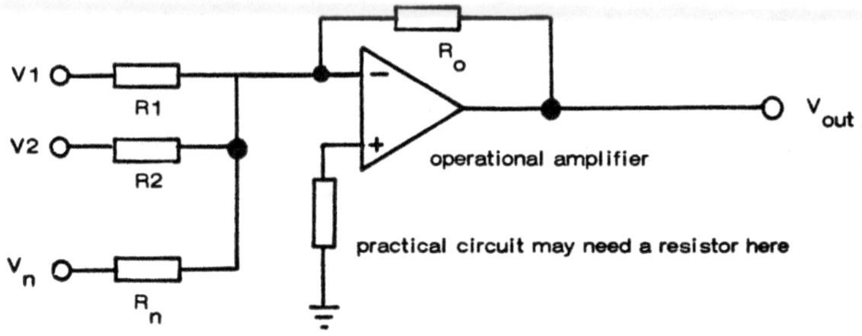

3.2 Adder circuit based on an op-amp. This inverts the output.

The specific performance of the op-amp as an adder can be calculated from:

$$Vout = - R_0 \; \frac{V1}{R_1} + \frac{V2}{R_2} \; \ldots + \frac{VN}{R_n}$$

Note the minus sign. This means that the output is 'inverted', i.e. this circuit shows an *inverting adder*.

By changing the inputs + to −, the op-amp will work as a *non-inverting adder*—Fig. 3.3.

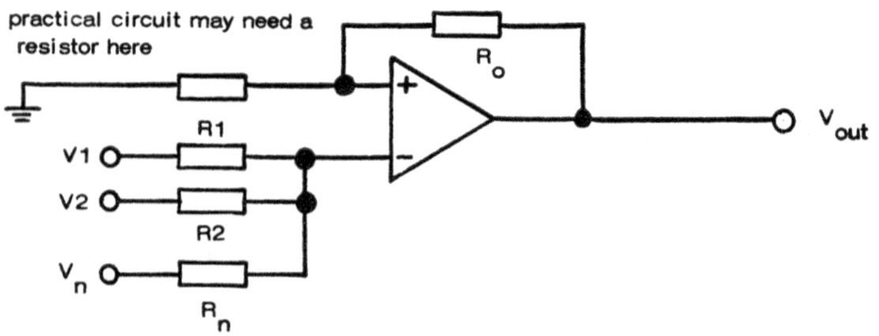

3.3 Non-inverting adder circuit, i.e. the input and output have the same polarity of signal and are thus in phase.

Amplifier or Buffer

Fig. 3.4 shows the circuit for an inverted amplifier or *inverter*. The gain is equal to -R2/R1, i.e. if R1 = R2 the

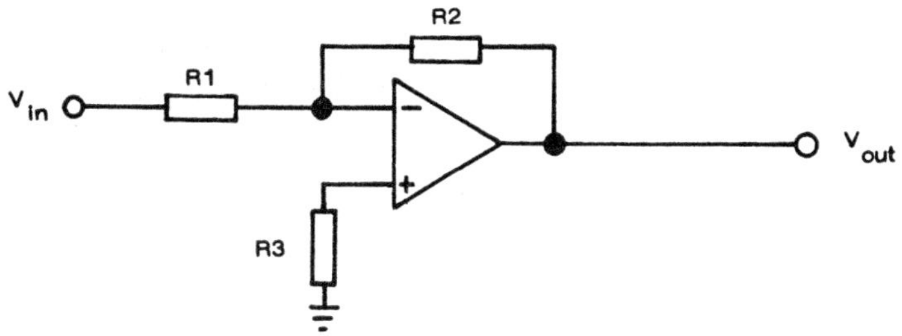

3.4 A circuit which inverts the input signal, known as an inverter. Voltage gain is R2/R1.

voltage gain is -1, meaning that the circuit works as a *phase inverter*. A suitable value for R3 can be calculated from:

$$R3 = \frac{R1 \times R2}{R1 \times R2}$$

To work as a non-inverting buffer (amplifier), the connections are made as shown in Fig. 3.5. In this case the gain is given by:

$$\text{gain} = 1 + \frac{R2}{R1}$$

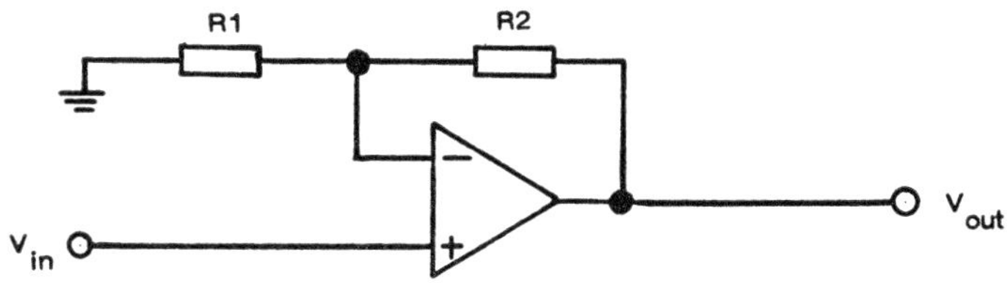

3.5 Non-inverting amplifier circuit where the gain is equal to 1 + R2/R1. If the gain is unity, this is also known as a buffer.

Multiplier

This is the same circuit as Fig. 3.4, using precision resistors of the specified values for R1 and R2 to give an exactly constant gain (and thus multiplication of input voltage in the ratio R2/R1). Note that this circuit inverts the phase of the output.

Adder/Subtractor

Connections for an adder/subtractor circuit are shown in Fig. 3.6. If R1 and R2 are the same value; and R3 and R4 are also made the same value as each other, then:

$$V\,out = V3 + V4 - V1 - V2$$

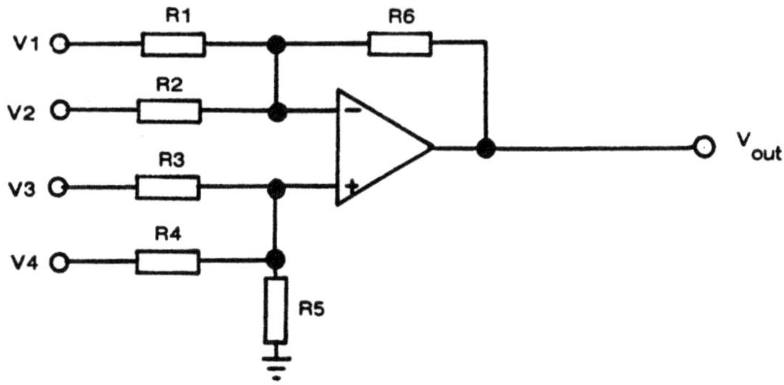

3.6 Adder/subtractor circuit. *See* text for calculation of component values.

In other words, inputs to V3 and V4 give a summed output (V out = V3 + V4). Inputs V1 and V2 subtract from the output voltage.

Values for R1, R2, and R3 and R4 are chosen to suit the op-amp characteristics. R5 should be the same value as R3 and R4; and R6 should be the same value as R1 and R2.

Integrator

Theoretically, at least, an op-amp will work as an *integrator* with the inverting input connected to the output via a capacitor. In practice a resistor needs to be paralleled across this capacitor to provide *dc* stability as shown in Fig. 3.7.

This circuit integrates input signal with the following relationship applying:

$$V\,out = \frac{1}{R1.C}\int V\,in\,dt$$

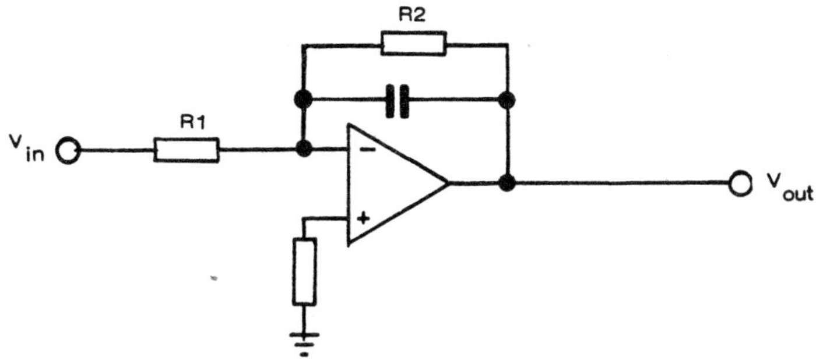

3.7 Op-amp integrator circuit.

The value of R2 should be chosen to match the op-amp characteristics so that:

$$V_{oo} = \frac{R2}{R1} \cdot V_{io}$$

Differentiator

The differentiator circuit has a capacitor in the input line connecting to the inverting input, and a resistor connecting this input to output. Again this circuit has practical limitations, so a better configuration is to parallel the resistor with a capacitor as shown in Fig. 3.8.

3.8 Practical circuit for an op-amp differentiator.

The performance of this circuit is given by:

$$V\,out = -\,R2C1\frac{dV\,in}{dt}$$

Differential Amplifier

A basic circuit for a differential amplifier is shown in Fig. 3.9. Component values are chosen so that R1 = R2 and R3 = R4. Performance is then given by:

$$V\,out = V\,in\,2 - V\,in\,1$$

provided the op-amp used can accept the fact that the impedance for input 1 and input 2 is different (impedance for input 1 = R1; and impedance for input 2 = R1 + R3).

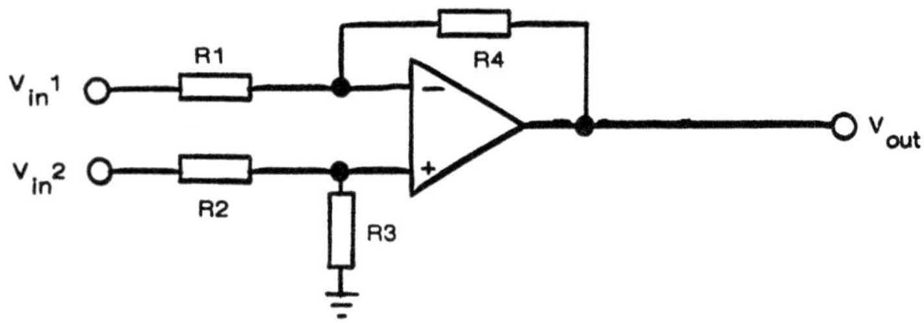

3.9 Basic differential amplifier circuit.

Log Amplifiers

The basic circuit (Fig. 3.10) uses an NPN transistor in conjunction with an op amp to produce an output proportional to the log of the input:

$$V\,out = -\,k\,\log_{10}\frac{Vin}{RI_o}$$

The right hand diagram shows the 'inverted' circuit, this

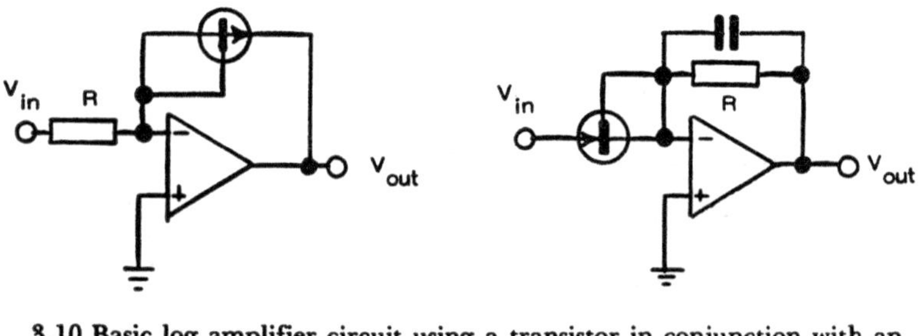

3.10 Basic log amplifier circuit using a transistor in conjunction with an op-amp.

time using a PNP transistor, to work as a basic *anti-log* amplifier.

The capacitor required is usually of small value (e.g. 20pF).

Log Multiplier

Logarithmic working of an op-amp is extended in Fig. 3.11 to give a *log multiplier*. Input X to one log amplifier gives *log X* output; and input Y to the second log amplifier gives *log Y* output. These are fed as inputs to a third op-amp to give an output *log XY*.

If this output is fed to an anti-log amplifier, the output is the inverted product of X and Y (i.e. *X.Y*).

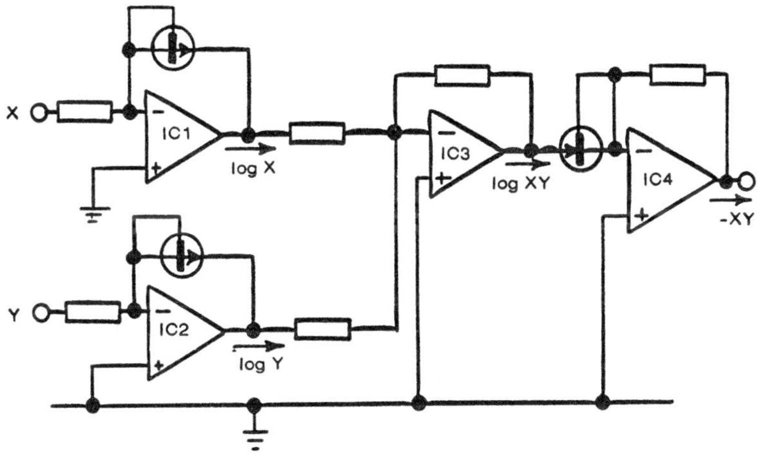

3.11 This circuit provides logarithmic working (*see* text).

Voltage Follower

Because of the inherent characteristics of an op-amp the connections shown in Fig. 3.12 will tie the two inputs together so that the output always follows the input, i.e.

$$V\ out = V\ in$$

The value of such a voltage follower is that it offers high input resistance with low input current and very low output resistance. There are many practical applications of this type of circuit and a number of op-amps are designed specifically as voltage followers.

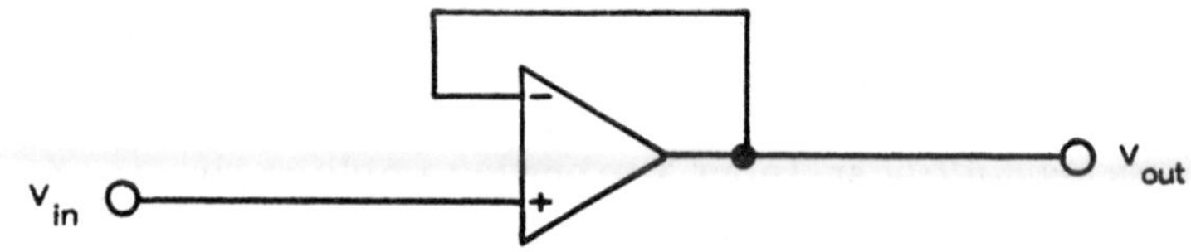

3.12 Tying the two inputs of an op-amp together gives a voltage-follower circuit where V out = V in. A characteristic of this circuit is high input resistance and very low output resistance.

Voltage-to-Current Converter

The circuit configuration shown in Fig. 3.13 will result in the same current flowing through R1 and the load impedance R2,

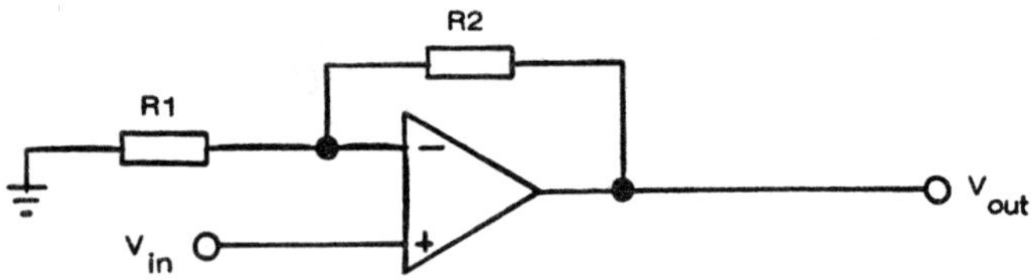

3.13 Voltage-to-current converter using an op-amp.

the value of this current being independent of the load and proportional to the signal voltage, although it will be of relatively low value because of the high input resistance presented by the non-inverting terminal. The value of this current is directly proportional to V in/R1.

Current-to-Voltage Converter

This configuration (Fig. 3.14) enables the input signal current to flow directly through the feedback resistor R2 when the output voltage is equal to Iin × R2. In other words, input current is converted into a proportional output voltage. No current flows through R2, the lower limit of current flow being established by the bias circuit generated at the inverting input.

A capacitor may be added to this circuit, as shown in the diagram, to reduce 'noise'.

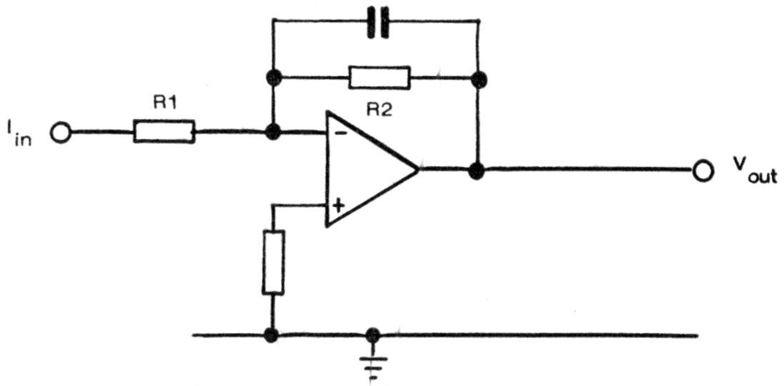

3.14 Current-to-voltage converter using an op-amp.

Current Source

Use of an op-amp as a current source is shown in Fig. 3.15.

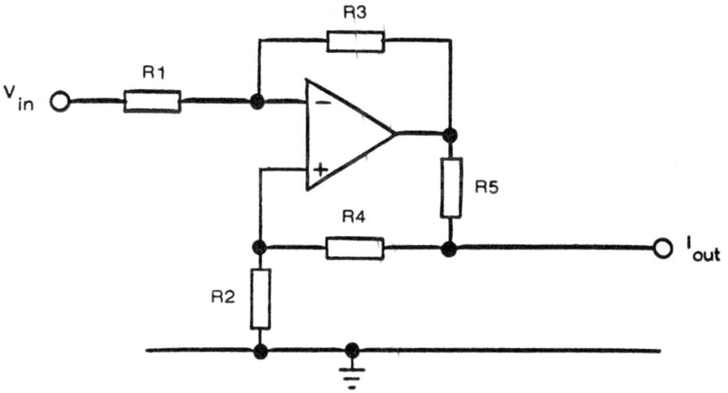

3.15 Circuit for using an op-amp as a current source. *See* text for component values required.

Resistor values are selected as follows:

$$R1 = R2$$
$$R3 = R4 + R5$$

Current output is given by:

$$I\ out = \frac{R3\ .\ Vin}{R1\ .\ R5}$$

Multivibrator

An op-amp can be made to work as a multivibrator. Two basic circuits are shown in Fig. 3.16. The one on the left is a free

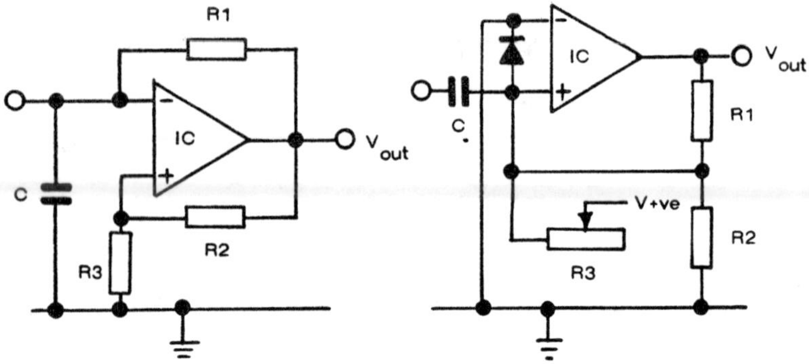

3.16 Two basic circuits for a multivibrator, based on op-amps. Component values for the right hand circuit are:

R1 – 1 Mohm
R2 – 10 Mohm potentiometer
R3 – 2 kohm
R4 – 1 kohm
diode – silicon diode
IC – CA741

C1 – 470 pF
C2 – 0.1 μF

running (astable) multi-vibrator, the frequency of which is determined by:

$$f = \frac{1}{2C.R1 \log_e \frac{2R3+1}{R2}}$$

The right hand diagram shows a monostable multivibrator circuit which can be triggered by a square wave pulse input. Component values given are for a CA741 op-amp.
See also separate chapter on 'Multivibrators.'

Schmitt Trigger

A *Schmitt trigger* is known technically as a *regenerative comparator*. Its main use is to convert a slowly varying input voltage into an output signal at a precise value of input voltage. In other words it acts as a voltage 'trigger' with a 'backlash' feature, called hysteresis.

The op-amp is a simple basis for a Schmitt trigger – *see* Fig. 3.17. The triggering or *trip* voltage is determined by:

$$V\,trip = -\frac{V\,out\,.\,R1}{R1 + R2}$$

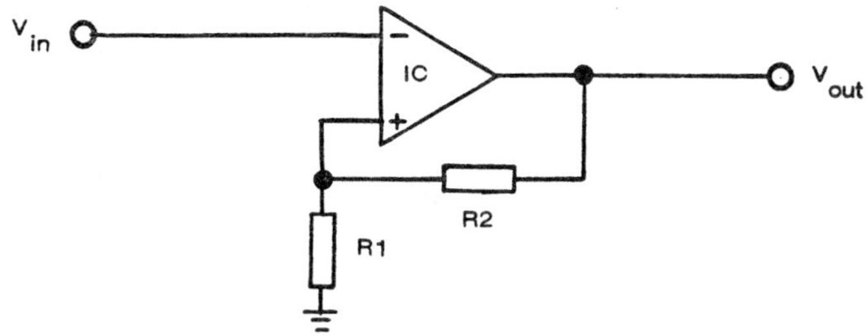

3.17 Schmitt trigger which gives an output once a precise value of varying input voltage is reached. An application of this circuit is a *dc* voltage level senser.

The hysteresis of such a circuit is twice the trip voltage.

Another Schmitt trigger circuit is shown in Fig. 3.18, the triggering point being approximately one-fifth of the supply voltage, i.e. there is a 'triggered' output once the *dc* input reaches one-fifth the value of the supply voltage. The supply voltage can range from 6 to 15 volts, thus the trigger can be

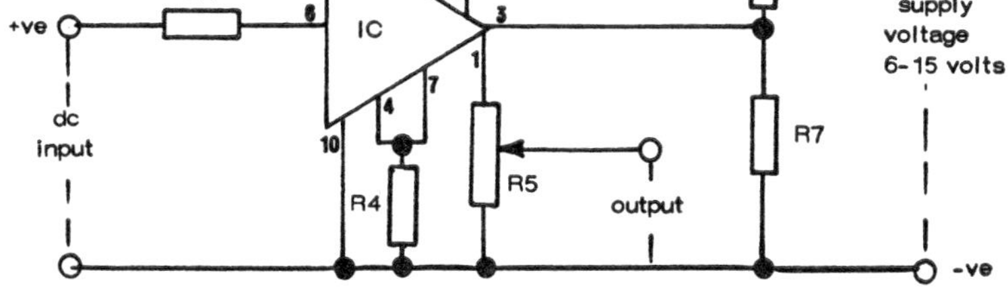

3.18 A more complicated Schmitt trigger circuit for general use.
Component values:
R1—2.2k ohm
R2—3.3k ohm
R3—3.3k ohm
R4—470 ohms
R5—5k ohm potentiometer
R6—33kohms
R7—22kohms
IC—CA3018

made to work at anything from 1.2 to 3 volts, depending on the supply voltage used. The actual triggering point can also be adjusted by using different values for R4, if required.

Once triggered, the output will be equal to that of the supply voltage. If output is connected to a filament bulb or LED (with ballast resistor in series), the bulb (or LED) will light once the input voltage has risen to the triggering voltage and thus indicate that this specific voltage level has been reached at the input.

Capacitance Booster

The circuit shown in Fig. 3.19 works as a *multiplier* for the capacitor C1, i.e. associated with a fixed value of C1 it gives an

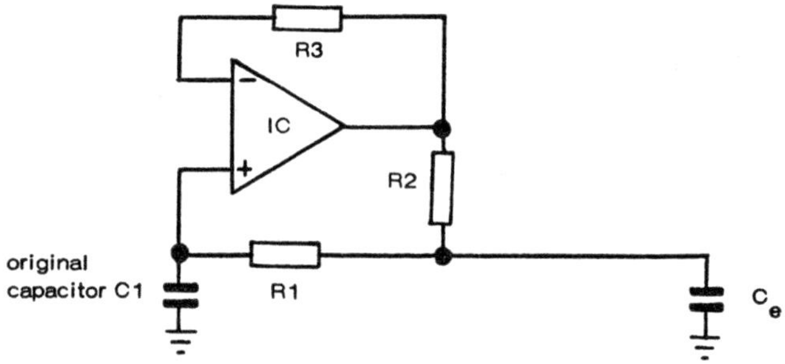

3.19 Capacitance multiplier circuit. The effective capacitance Ce is equal to the value of C1 multiplied by R1/R2.

effective capacitance Ce which can be many times greater. The actual multiplication ratio is R1/R2 so that making R1 ten times greater than R2, say, means that the effective capacitance of this circuit would be 10 × C1.

As far as utilization of such a multiplier is concerned, the circuit now also contains resistance (R2) in series with the effective capacitance.

Filters

Op-amps are widely used as basic components in filter circuits. Two basic circuits are shown in Fig. 3.20. The one on the left is a *low pass filter* and the one on the right is a *high pass filter*.

See also separate chapter on Filters.

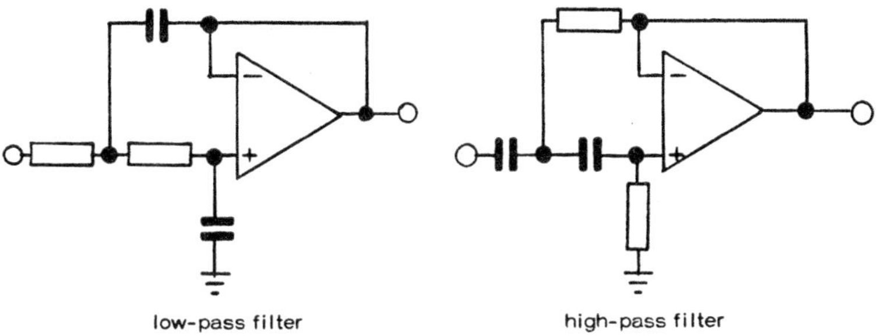

3.20 Two basic filter circuits using op-amps.

Op-amp Parameters

The ideal op-amp is perfectly balanced so that if fed with equal inputs, output is zero, i.e.

$$V\ in\ 1 = V\ in\ 2 \text{ gives } V\ out = 0$$

In a practical op-amp the input is not perfectly balanced so that unequal bias currents flow through the input terminal. Thus an input offset voltage must be applied between the two input terminals to balance the amplifier output.

The *input bias current* (I_B) is one half the sum of the separate currents entering the two input terminals when the output is balanced, i.e. V out = 0. It is usually a small value, e.g. a typical value is $I_B = 100nA$.

The *input offset current* (I_{io}) is the difference between the separate currents entering the input terminals. Again it is usually of a very small order, e.g. a typical value is $I_{io} = 10nA$.

The *input offset voltage* (V_{io}) is a voltage which must be applied across the input terminal, to balance the amplifier. Typical value, $V_{io} = 1mV$.

Both I_{io} and V_{io} are subject to change with temperature, this change being known as I_{io} drift and V_{io} drift, espectively.

The *Power Supply Rejection Ratio* (PSRR) is the ratio of the change in input offset voltage to the corresponding change in one power supply voltage. Typically this is of the order of $10-20\mu V/V$

Other parameters which may be quoted for op-amps are:

Open-loop gain—usually designated A_d.

Common-mode rejection ratio—designated CMPR or ρ. This is the ratio of the difference signal to the common-mode signal and represents a figure of merit for a differential amplifier. This ratio is expressed in decibels (dB).

Slew rate—or the rate of change of amplifier output voltage under large—signal conditions. It is expressed in terms of V/μs.

Chapter Four

AUDIO AMPLIFIERS

QUITE A NUMBER of linear ICs are designed as audio amplifiers for use in radio receivers, record players, etc. Again these are used with external components but physical layout, and the length of leads is relatively unimportant—unlike circuits carrying radio frequencies. The 'packaging' of such ICs can vary from cans (usually TO-5 to TO-100 configuration) to dual-in-line and quad-in-line. In all cases they will usually have 12 or 14 leads (but sometimes less). Not all these leads are necessarily used in a working circuit. They are there to provide access to different parts of the integrated circuit for different applications. Integrated circuits designed with higher power ratings may also incorporate a tab or tabs to be connected to a heat sink; or a copper slug on top of the package for a similar purpose.

A single chip can contain one, two, three or more amplifier

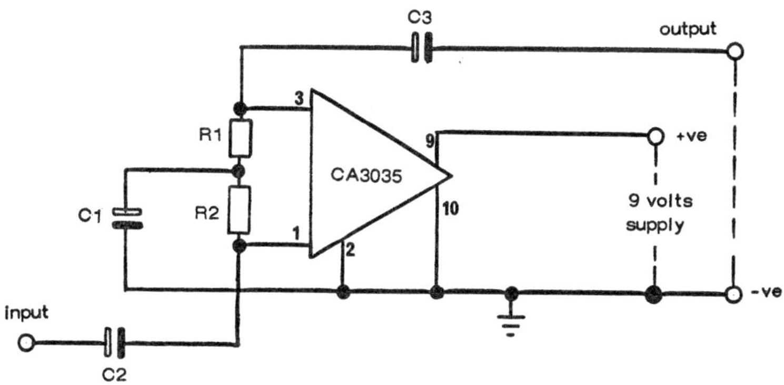

4.1 Utilization of the first amplifier in CA3035 integrated circuit by tapping pins 1, 2, 3, 9 and 10. This circuit gives a voltage gain of 100-160 with an input resistance of 50 k ohms and an output resistance of 270 ohms.

Component values:

R1 — 100kohms	C1 — 10μF
R2 — 100kohms	C3 — 1μF
	C3 — 10μF

stages interconnected and following each other (technically referred to as being in *cascade*). Pin-out connections provide 'tapping' points for using one or more stages separately or in cascade as required.

The (RCA) CA3035 integrated circuit is just one example. It consists of three separate amplifier stages connected in cascade with a component count equivalent to 10 transistors, 1 diode and 15 resistors. Each amplifier stage has different characteristics. The first stage, which can be selected by connections to pins 1, 2, 3, 9 and 10 (*see* Fig. 4.1), is a wide band amplifier characterized by high input resistance (i.e. ideally suited to connecting to a preceding transistor stage). The working circuit using this stage is shown in Fig. 4.1. It has a gain of the order of 160 (44dB).

The second amplifier in the CA3035 has a lower input resistance (2k ohm) and a low output resistance of 170 ohms. The gain is similar to the first stage (about 45dB). A working circuit with tapping points is shown in Fig. 4.2.

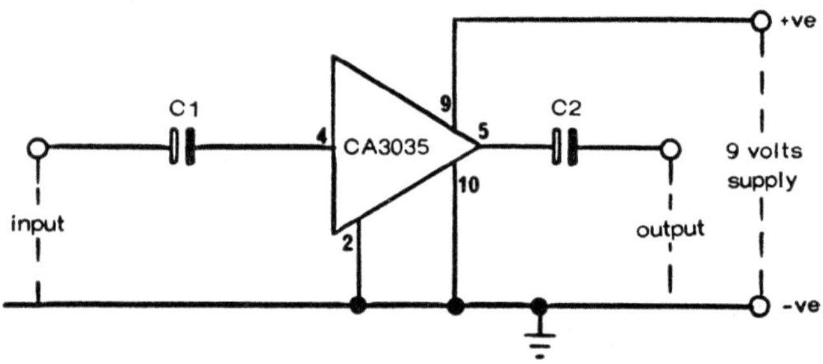

4.2 Utilization of the second amplifier in CA3035 integrated circuit by tapping pins 2, 4, 5, 9 and 10. This circuit gives a voltage gain of 100-120 with an input resistance of 2 k ohms and an output resistance of 170 ohms.
Component values:
C1 – 10μF
C2 – 10μF

The third amplifier is a wide band amplifier with a low input resistance (670 ohms) and a high output resistance (5k ohms). It offers a voltage gain of 100 (40dB). A working circuit is shown in Fig. 4.3.

Amplifiers 1 and 2 can be cascaded; or amplifiers 2 and 3; or

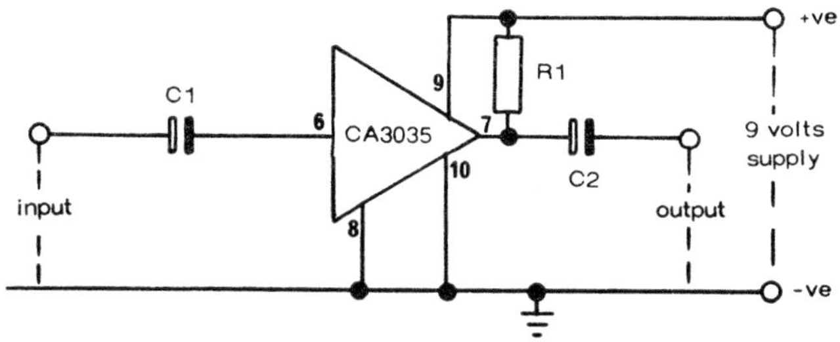

4.3 Utilization of the third amplifier in CA3035 integrated circuit by tapping pins 6, 7, 8, 9 and 10. This circuit gives a voltage gain of 80-120 with an input resistance of 670 ohms and an output resistance of 5 k ohms.

Component values:

R1 – 5 kohms
C1 – 10μF
C2 – 10μF

amplifiers 1, 2 and 3. Fig. 4.4 shows the external connections and components required to cascade amplifiers 1 and 2.

Using all three amplifiers in cascade results in a gain of approximately 110 dB. The circuit in this case is shown in Fig. 4.5.

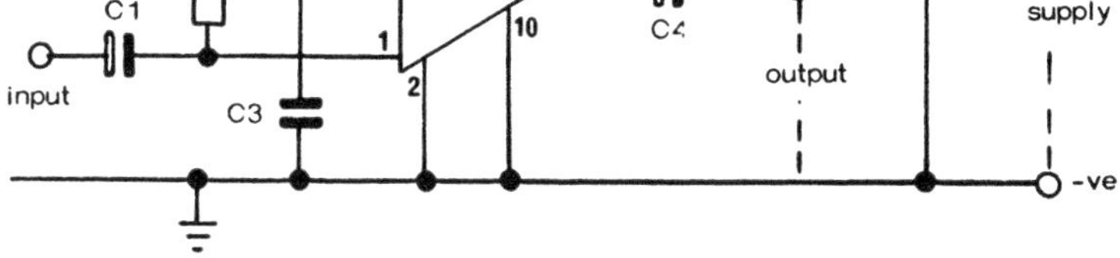

4.4 Circuit for using first and second amplifiers contained in CA3035 in cascade. This circuit gives a voltage gain of about 7000 with an input of 50 k ohms and an output resistance of 170 ohms.

Component values:

R1 – 220 kohms
R2 – 1.2 kohms

C1 – 10μF
C2 – 0.22μF
C3 – 0.04μF
C4 – 10μF
C5 – 50μF

Modifying Amplifier Performance

The output impedance of an amplifier stage can be modified by connecting R1 to provide a negative feedback from output to input. This has the effect of reducing the working value of R1 and R1/Av where Av is the amplifier open loop voltage gain. This is accomplished without affecting the actual voltage gain. In the case of cascaded amplifiers a capacitor C2 is needed in series with R1 to act as a block to *dc* (i.e. R1 only is needed for amplifier 1 part of CA 3035, C1 being effective as a blocking capacitor in this case). Since amplifier 2 in this chip is directly

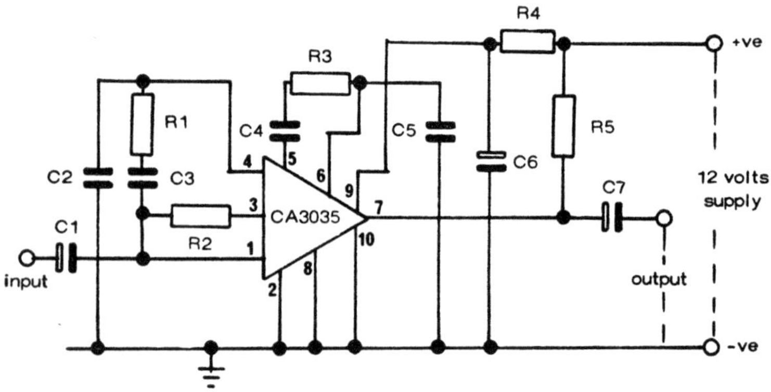

4.5 This circuit shows all three amplifiers in CA3035 cascaded to give a voltage gain of about 200,000

Component values:

R1 — 220 kohms	C1 — 10 μF
R2 — 1.2 kohms	C2 — 0.04 μF
R3 — 680 ohms	C3 — 0.22 μF
R4 — 1 kohm	C4 — 0.05 μF
R5 — 4.7 kohms	C5 — 0.05 μF
	C6 — 50 μF
	C7 — 10 μF

coupled to amplifier 1; and amplifier 2 is directly coupled to amplifier 3; the use of an impedance-matching resistor applied to amplifier 2 (or amplifier 3) will require the use of a blocking capacitor in series with the resistor.

The gain of the amplifier stage can be modified by the use of a series resistor in the input (R1). This acts as a potential divider in conjunction with the effective input resistance of the stage so that only a proportion of the input signal is applied to the stage. In this case:

1. actual voltage gain $= \dfrac{R1}{Ri + R1/Av}$

2. input resistance $= Ri + R1/Av$

where Ri is the input resistance of the IC

Thus by suitable choice of R1 and Ri, both voltage gain and input resistance of an amplifier circuit can be modified to match specific requirements. It follows that if a number of different resistors are used for Ri, the circuit can be given different response (sensitivity) for a given input applied to each value of Ri by switching. This mode of working is useful for pre-amplifiers. Virtually the same circuit is used for an audio mixer, separate input channels being connected by separate series resistors (Ri) and thence commonly connected to the

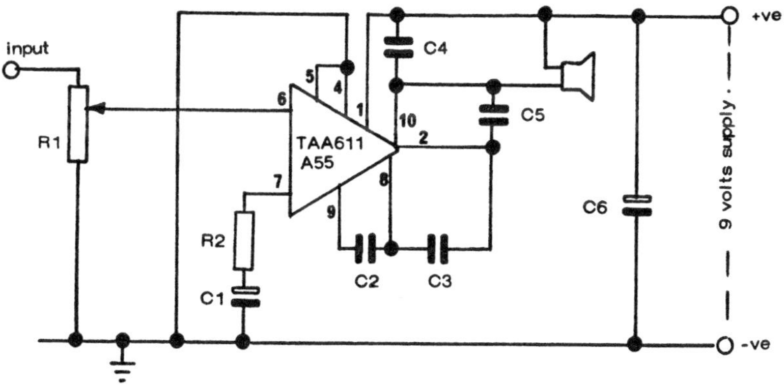

4.6 Audio amplifier for radio receiver based on the TAA611 A55 integrated circuit. Pin numbers shown are for the can-shaped version of this IC.

Component values:

R1 — 22 kohms
R2 — 30 ohms
C1 — 50 μF/6 volt
C2 — 56 pF
C3 — 150 pF
C4 — 1 μF
C5 — 500 μF/12 volt
C6 — 100 μF/12 volt
loudspeaker — 8 ohms

input. In this case each channel has the same input resistance with an overall gain of unity.

Fig. 4.6 shows a circuit for a low power (1.8 watt) audio amplifier using a TA 611 monolithic integrated circuit. This particular IC is available in two configurations, a TO-100 metal case and in a quad-in-line plastic package. Lead positions are shown in Fig. 4.7 for the two different configurations.

This is a particularly attractive circuit for it needs a minimum number of external components and is capable of

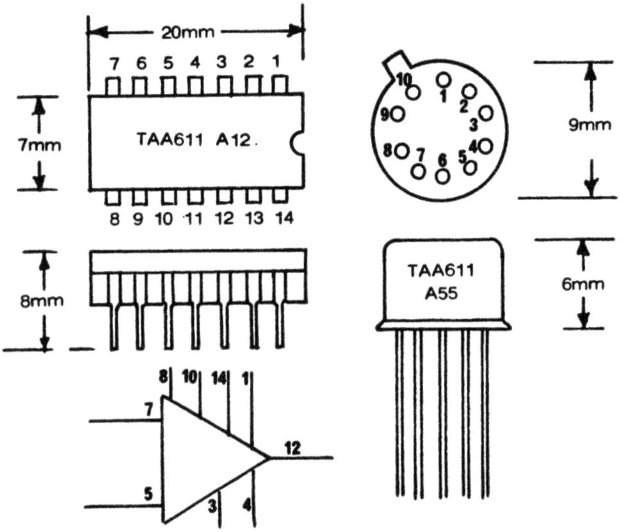

4.7 The two versions of the TAA611 integrated circuit. The TAA611 A55 is a 14-pin dual-in-line package. The TAA611 A55 is a 'can' shape package in a metal case (TO-100). The circuits are identical so either can be used in Fig. 4.1 with the same external components. Note, however, the different pin-out arrangement for the TAA611 A12 on the left.

driving an 8-ohm loudspeaker direct with any supply voltage between 6 volts and 12 volts. Also it does not require a heat sink.

Exactly the same circuit can be used with a number of other integrated circuits in the same family, offering higher output powers. These are the TA611B and TA611C. The only difference is the values required for the external components required, viz:

	TA611B supply volts 6 - 15	TA611C supply volts 6 - 18
max. power output	2.1 watts	3.3 watts
R1	22k ohms	220k ohms
R2	30 ohms	150 ohms
C1	50μF	25μF
C2	56pF	82pF
C3	150pF	1.2μF
C4	omit	omit
C5	500μF	500μF
C6	100μF	100μF

Lead arrangement for the TA611B and TA611C are the same as TAA611 A12.

Because of its higher power the current based on the TA611C really requires the IC to be mounted with a heat sink (Chapter 5 deals specifically with heat sinks), although this is not absolutely essential. The type is, in fact, available with a special mounting bar or spacer to which a heat sink can be attached. The recommended method of mounting is shown in Fig. 4.8, the heat sink itself being a piece of aluminium sheet cut to a suitable size and bent to the shape shown. The IC itself has a copper slug on its top face on to which the heat sink sits (and is clamped down by the mounting bolts). Better thermal contact between the IC and the heat sink can be achieved if the contact area is very lightly coated with silicon grease.

There are other methods of fitting heat sinks to this IC (and other types). The TA611C is also available with an external bar, the ends of which can be soldered to copper 'patches' on the printed circuit panel (also shown in Fig. 4.8). In this example the copper areas form the actual heat sink. A suitable area in this case would be about 30 mm square for each copper patch. These copper areas are, of course, merely used for heat dissipation and are not part of the actual printed circuit as such, although it is normally advisable—and necessary with some types of IC—to connect the heat sink area to the common earth of the circuit. It is just a convenient method of making heat sinks integral with (and at the same time as) the printed circuit panel.

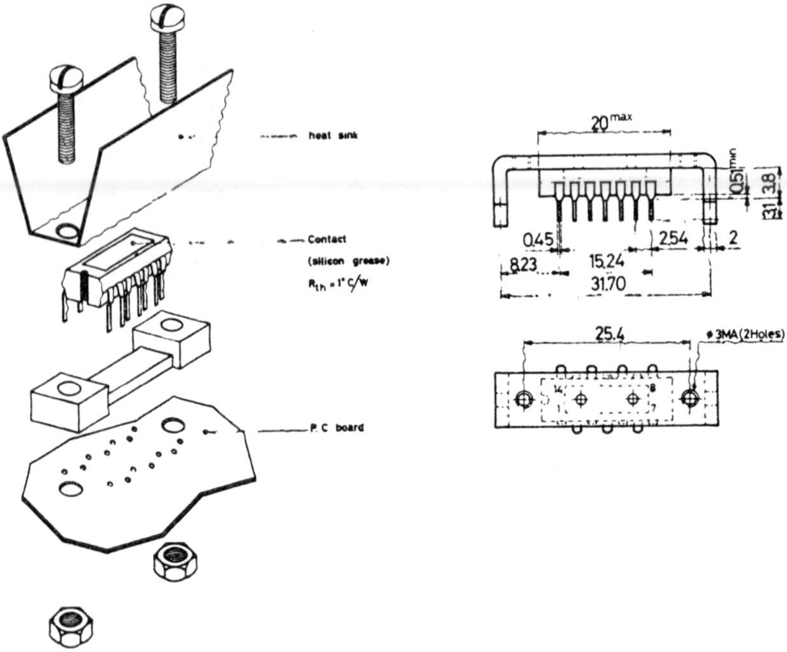

4.8 Aluminium sheet heat sink applied to the TA611C integrated circuit (*left*) and alternative external bar fitted to this IC (*right*) for connecting to heat sink areas on copper of printed circuit board.

A further audio amplifier based on a TBA641B integrated circuit is shown in Fig. 4.9. This is a little more complicated in terms of the number of external components used but has the advantage of driving a 4-ohm speaker (the more readily available value with larger loudspeakers) and is suitable for direct coupling of the input. It will operate on a supply voltage ranging from 6 volts to 16 volts and give 4.5 watts output power at 14 volts. Again the IC needs mounting with a heat sink of the type illustrated in Fig. 4.8.

Short Circuit Protection

A feature of many audio amplifier circuits is that if the output is shorted when the circuit is switched on (e.g. loudspeaker connections accidentally shorted), excessive current may be passed by the output transistors sufficient to destroy

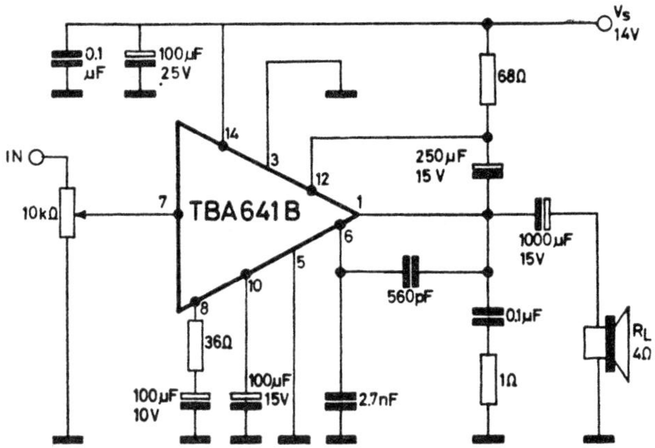

4.9 Audio amplifier for 4-ohms loudspeaker based on the TBA641B integrated circuit. Component values are shown on the diagram (SGS-Gates).

them. It is possible to provide short-circuit protection with additional circuitry limiting the current which can flow through the output transistors. This can readily be incorporated in an IC, an example being the TCA940 designed as a 10 watt class B amplifier. Other characteristics of this particular IC are high output current (up to 3 amps) very low harmonic and crossover distortion and a thermal shut down feature (See later).

The circuit is shown in Fig. 4.10. Supply voltage is up to 24 volts. Power rating depends both on the supply voltage used and the resistance of the loudspeaker:

Supply voltage	4-ohm speaker	8-ohm speaker
20	10 watts	6.5 watts
18	9 watts	5 watts
16	7 watts	

A feature of this circuit is that the bandwidth is controlled by the values of Rf and C3 and C7. For a value of Rf = 56 ohms with C3 = 1000pF and C7 = 5000pf the bandwidth is 20kHz. For the same *capacitor* values the bandwidth can be reduced to 10kHz by making Rf = 20 ohms. For the original resistor value (Rf = 56 ohms), the bandwidth can be reduced to 10kHz by making C3 = 2000pf and C7 = 10000pF.

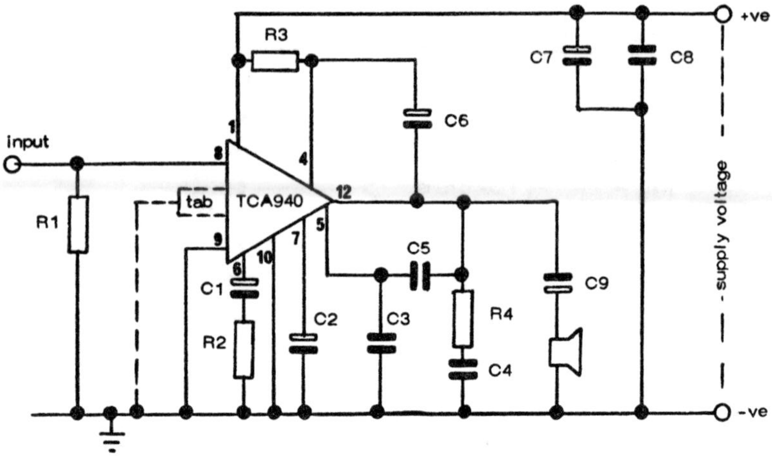

4.10 10-watt amplifier circuit based on the TCA940 integrated circuit. The TCA940 is a 12-lead quad-in-line plastic package.

Component values:

R1 — 100 kohms
R2 — 56 ohms
R3 — 100 ohms
R4 — 1 ohm
C1 — 100 μF/3 volt
C2 — 100 μF/15 volt
C3 — 4700 pF
C4 — 1 μF
C5 — 1000 pF
C6 — 100 μF/15 volts
C7 — 100 μF/25 volts
C8 — 0.1 μF
C9 — 2000 μF/16 volts

Circuit assembly is straightforward, except that the IC needs a heat sink. It is provided with tabs which should be bolted to an external aluminium sheet heat sink of generous area.

Thermal Shut Down

The short-circuit protection built into this IC effectively works as a power-limiting device. It is only effective on a short-duration basis, i.e. to provide protection against temporary overload and short circuiting of the output. An additional circuit is included to ensure that regardless of how long a short circuit is present across the output the junction temperature of the output transistors is kept within safe limits.

In other words, this additional piece of circuitry incorporated in the IC provides complete protection against a shorted output. It also has another advantage. The same protection is present if there is another cause of overheating, e.g. the heat sink used is not really large enough for the job it is intended to do. The thermal shut-down circuitry simply reacts to the junction temperature becoming too high by reducing the output current and power to compensate.

Hi-Fi Stereo Amplifier

The excellent performance and extremely good stability possible with integrated circuits makes them a logical choice for Hi-Fi circuits. The TDA 2020 monolithic integrated operational amplifier is an excellent up-to-date example of such a device, designed to be used as a Class B audio power amplifier for Hi-Fi circuits. It is capable of providing a 20-watt power output into a 4-ohm loudspeaker with a supply voltage of 18 volts, and with a guaranteed output power of 15 watts. It is also a device for providing high output current (up to 3.5 amps) and has a very low harmonic and cross-over distortion. It also incorporates short-circuit protection and thermal shut-down protection.

The TDA2020 is in the form of a quad-in-line plastic package of conventional appearance with 14 leads. Because of its high power rating it is intended to be used with a specially formed heat sink mounted on a spacer designed to provide proper thermal contact between the IC itself and the heatsink when assembled on two bolts—Fig. 4.11. The most negative supply voltage of the circuit is connected to the copper slug on the IC and hence also to the heat sink.

The basic amplifier circuit is completed by the addition of four external resistors and seven capacitors, plus a coupling capacitor to enable the circuit to be used with a split power supply. This provides direct drive for a 4-ohm loudspeaker. Since the Hi-Fi circuits are usually stereo, two ICs are used in this basic circuit configuration, each IC powering its own loudspeaker. The complete circuit is shown in Fig. 4.12.

Another simpler stereo audio amplifier circuit is shown in Fig. 4.13, based on the (Mullard) TDA1009 integrated circuit. This IC is a low frequency Class B amplifier with no crossover

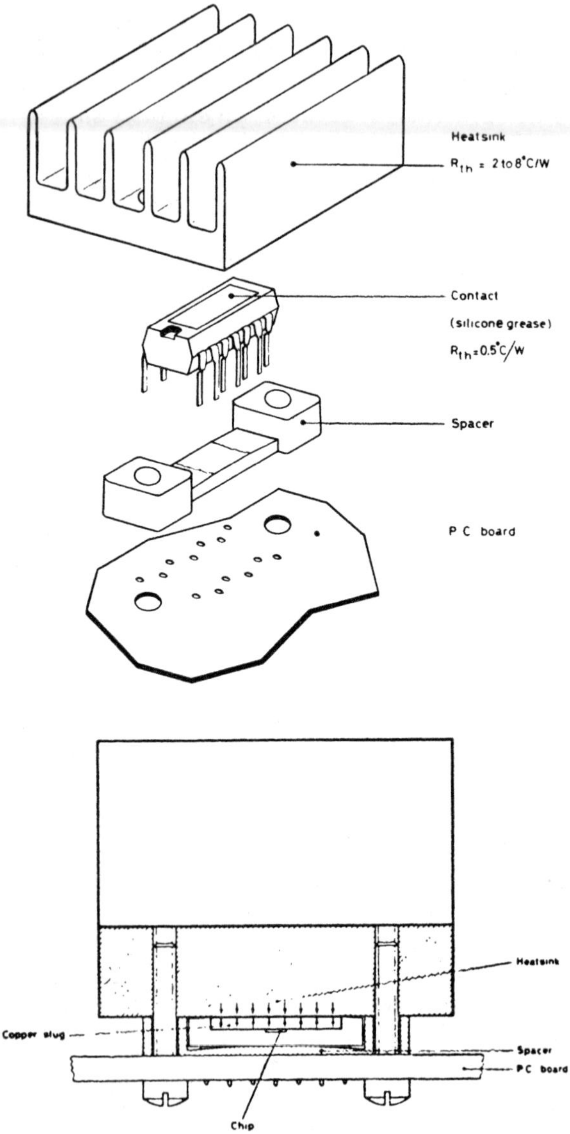

4.11 Heat sink for TDA2020 integrated circuit.

distortion designed for use with a minimum number of external components. It delivers 2 x 6 watts output power at 10 per cent distortion into speakers of 4-ohms impedance with 8 to 16 volts supply; and can also deliver the same power into speakers with 8-ohms impedance using a 24-volt supply. The IC

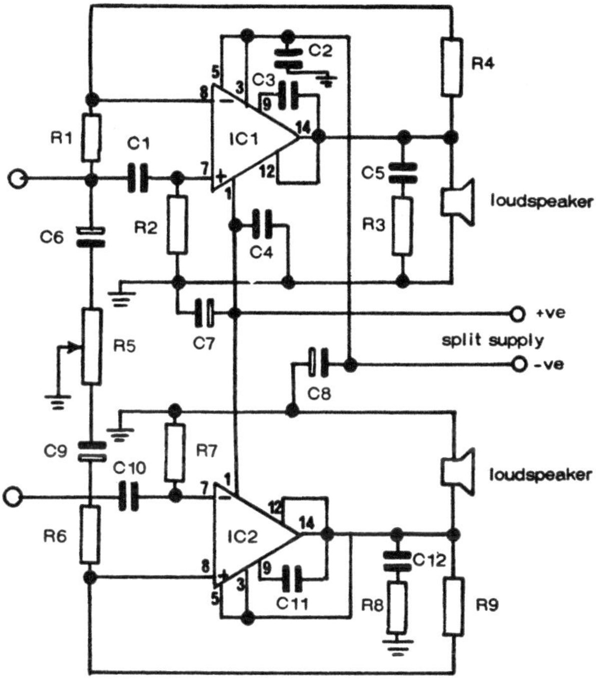

4.12 Stereo amplifier circuit with split supply voltage ±17 to ±24 volts.
Component values:

R1 – 1 kohm	C1 – 0.1 μF
R2 – 100 kohms	C2 – 0.1 μF
R3 – 1 ohm	C3 – 68 pF
R4 – 100 kohms	C4 – 0.1 μF
R5 – 47 kohms	C5 – 0.1 μF
R6 – 1 kohm	C6 – 47 μF
R7 – 100 kohms	C7 – 100 μF
R8 – 1 ohm	C8 – 100 μF
R9 – 100 kohms	C9 – 47 μF
IC1 – TDA2020	C10 – 0.1 μF
IC2 – TDA2020	C11 – 68 pF
loudspeakers – 4 ohms	C12 – 0.1 μF

incorporates short circuit protection for supply voltages up to 16 volts and also thermal protection. Input impedance is 45k ohms.

The addition of capacitors C9 and C10 (shown dotted) provides 'bootstrapping'. This provides increased output power.

High Power Amplifiers

Most of the original IC audio amplifiers which appeared on the market had a relatively limited power output and thus

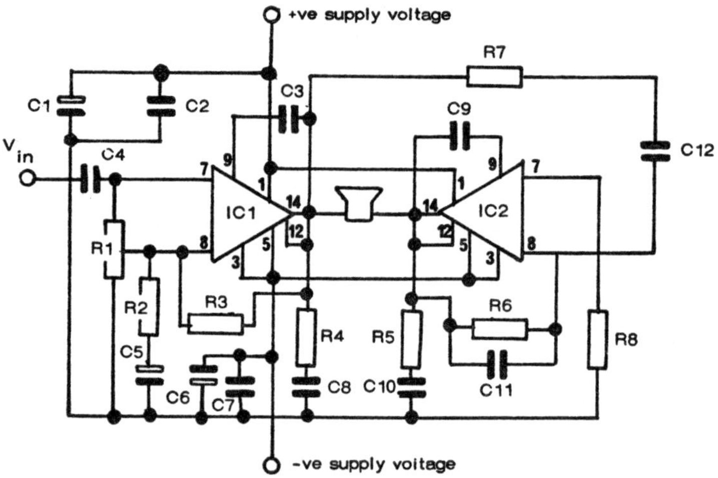

4.13 Bridge amplifier circuit with split power supply ±17 volts to ±22 volts.

Component values:

R1 – 100kohms	C1 – 100μF
R2 – 33kohms	C2 – 0.1μF
R3 – 100kohms	C3 – 68pF
R4 – 4ohms	C4 – 0.1μF
R5 – 1ohm	C5 – 4.7μF
R6 – 100kohms	C6 – 100μF
R7 – 100kohms	C7 – 0.1μF
R8 – 100kohms	C8 – 0.1μF
	C9 – 560pF
IC1 – TDA2020	C10 – 0.1μF
IC2 – TDA2020	C11 – 0.1μF
loudspeaker – 4 or 8ohms	C12 – 0.1μF

needed to be associated with a further stage or stages of transistor amplification to give more than a few watts output. Single IC amplifier chips are now readily available with output

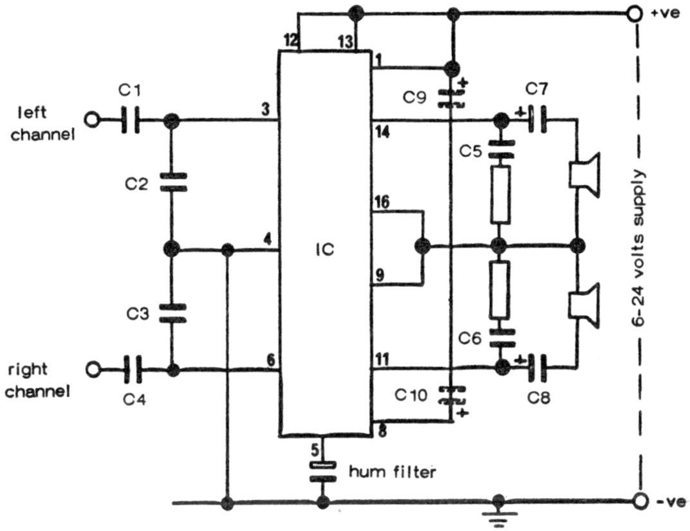

4.14 Simple 2 x 6 watts stereo amplifier circuit using a minimum of external components (Mullard).
Note the rectangular symbol used for the IC. This is often more convenient than a triangle or separate triangles when the integrated circuit has a large number of pins.
Component values:

Resistors
R1 —4.7ohms
R2 —4.7ohms
Capacitors
C1 —100mF
C2 —330pF
C3 —330pF
C4 —100mF
C5 —100mF
C6 —100mF
C7 —1000μF
C8 —1000μF
C9 *—47μF
C10*—47μF
*bootstrap capacitors
hum filter—47μF
IC—TDA1009

powers from 1 to 5 watts and substantially higher outputs are obtainable from later developments capable of handling even higher voltages and currents with satisfactory thermal stability.

Bridge Amplifiers

Even higher outputs are obtainable from *bridge amplifier* circuits. These can be used to increase power from output for a given supply voltage, or maintain a high power output with a reduced supply voltage. Bridge connection can give four times the output power under a given load with the same supply voltage; or twice the output power at a fixed peak current if the load impedance is doubled (e.g. using 8-ohm loudspeaker(s) instead of 4-ohm loudspeaker(s)).

An example of this type of circuitry, again using the TDA2020 integrated circuit is shown in Fig. 4.14. It is capable of delivering 30 watts power output in an 8-ohm loudspeaker with a supply voltage of ± 17 volts.

Chapter Five

HEAT SINKS

WHERE INTEGRATED CIRCUITS handle moderate powers and a heat sink is necessary to dissipate heat generated within the IC itself, areas etched on the copper of a printed circuit board can conveniently be used as heat sinks. ICs which are suitable for heat sinks of this type are usually fitted with a tab or tabs for soldering directly to the copper bands forming the heat sink.

The area of copper needed for a heat sink can be calculated quite simply, knowing the relevant circuit parameters. First it is necessary to determine the maximum power to be dissipated, using the formula:

$$\text{Power (watts)} = 0.4\,\frac{V_s^2}{8R_L} + V_s \,.\, I_d$$

where V_s is the maximum supply voltage
I_d is the quiescent drain current in amps under the most adverse conditions.
R_L is the load resistance (e.g. the loudspeaker resistance in the case of an audio amplifier circuit).

Strictly speaking the value of V_s used should be the battery voltage plus an additional 10 per cent, e.g. if the circuit is powered by a 12-volt battery, the value of V_s to use in the formula is $12 + 1.2 = 13.2$ volts. This allows for possible fluctuations in power level, such as when using a new battery. If the circuit has a stabilized power supply, then V_s can be taken as this supply voltage.

The quiescent drain current (I_d) is found from the IC parameters as specified by the manufacturers and will be dependent on supply voltage. Figures may be quoted for 'typical' and 'maximum'. In this case, use the maximum values.

Fig. 5.1 then gives the relationship between power to be dissipated and copper area, based on a maximum ambient temperature of 55°C (which is a safe limit for most IC devices). Example: Supply voltage for a particular IC is 12 volts. Load resistance is 4 ohms and the maximum quiescent current drain quoted for the IC at this operating voltage is 20 milliamps. The supply voltage is not stabilized, so the value to use for V_s is

$$12 + 1.2 = 13.2 \text{ volts}$$

$$\text{Thus power} = 0.4 \times \frac{13.2^2}{8 \times 4} + (13.2 \times 0.020)$$

$$= 2.178 + 0.264$$

$$= 2.422 \text{ watts (say 2.5 watts)}$$

From Fig. 5.1, a suitable copper area is seen to be a 40 mm square.

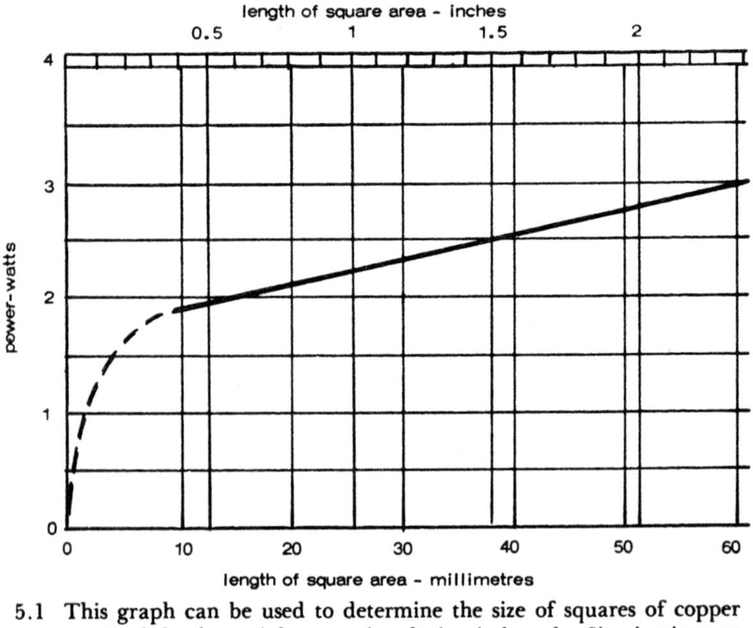

5.1 This graph can be used to determine the size of squares of copper required for heat sinks on printed circuit boards. Size is given as length of a square. Any other shape of the same area can be used (*see* text).

This heat sink area can be arranged in two squares (if the IC has two tags); or a single square (if the IC has one tag)—Fig. 5.2. Of course, the area does not have to be a square. This is simply the easiest shape to calculate. It can be rectangular, regular or irregular in shape, provided there is sufficient area. A point to be borne in mind, however, is that with any shape the copper area nearest the tag will have greater efficiency as a heat dissipator, so shapes which concentrate the area in this region are better than those that do not. If such a shape cannot

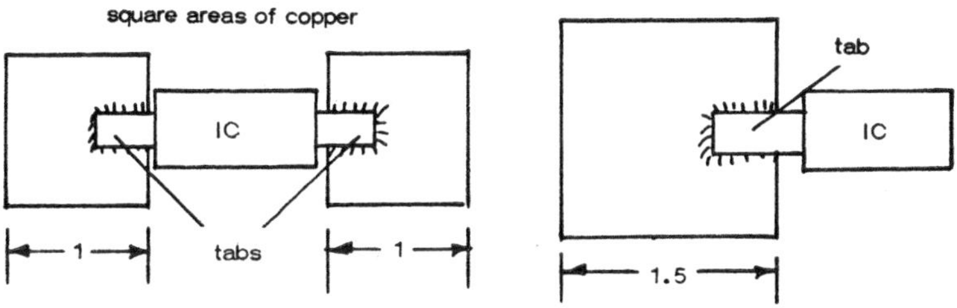

5.2 Copper area determined from Fig. 5.1 is for two equal squares (one at each end of the IC). If a single square is used at one end of the IC, its area needs to be slightly greater for the same heat dissipation.

be incorporated conveniently on the printed circuit layout a less efficient shape has to be used, then it may be necessary to increase the actual area of copper to compensate. Copper areas given in Fig. 5.1 should be adequate since most ICs can be worked at fairly ambient temperatures without trouble (e.g. up to 70°C). Very approximately this higher temperature operation will be provided by a copper area of a little more than one half that given by the graph, so there is a fair margin for error available when using this graph.

The graph also shows that the area of copper necessary to dissipate powers of 3 watts or more tends to become excessive, compared with the area of printed circuit panel actually required for the circuit itself. Where higher powers have to be dissipated, therefore, it is usually more convenient (and more efficient) to dissipate heat by an external heat sink fitted to the IC itself. Some examples of external heat sinks are shown in Figs. 4.8, 4.11 and 5.3.

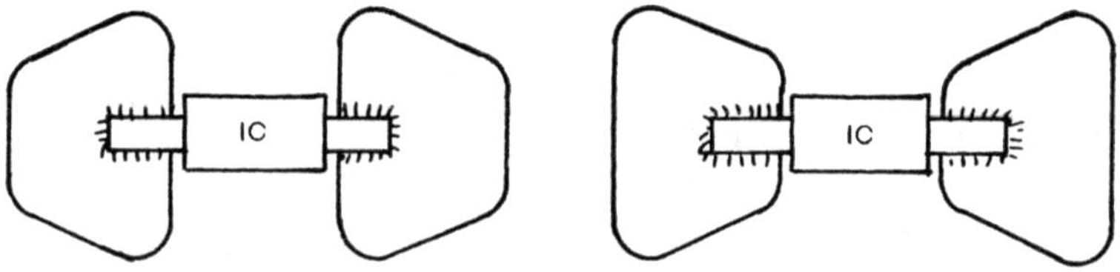

5.3 Good and poor shapes for heat sink areas on PCBs.

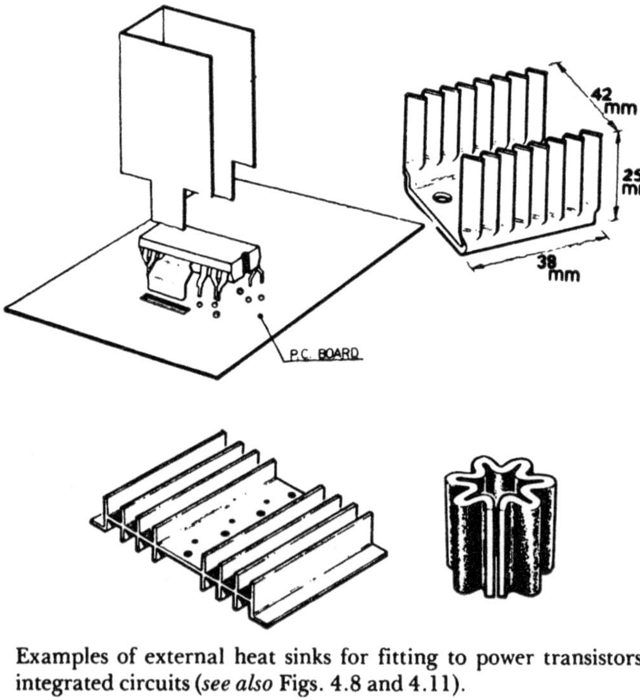

5.4 Examples of external heat sinks for fitting to power transistors and integrated circuits (*see also* Figs. 4.8 and 4.11).

Chapter Six

COMPLETE RADIO CIRCUITS

THE DEVELOPMENT OF RADIO CIRCUITS around a single IC with the same physical size (and shape) as a single transistor is exemplified by the following. The IC is the Ferranti ZN414 which contains the equivalent of 10 transistors in a complete TRF (tuned radio frequency) circuit providing three stages of RF application, a detector and *agc* (automatic gain control).

The ZN414 has three leads, identified as input, output and ground. It provides a complete radio circuit in itself to be connected to an external tuned circuit, an output decoupling capacitor, a feedback resistor and second decoupling capacitor, and an AGC resistor. As with any high gain RF device, certain requirements should be observed to ensure stable and reliable operation. These are:

1. All leads connecting components to the ICs should be kept as short as possible.
2. The output decoupling capacitor should be connected with very short leads to the output and ground leads of the ZN414.
3. The tuned circuit should be kept as far away as possible from the battery and from the loudspeaker and leads connecting these components to the circuit.
4. The 'earthy' side of the tuning capacitor (the moving part) must be connected to the junction of the feedback resistor and the second decoupling capacitor. A basic radio circuit using a minimum of components is shown in Fig. 6.1.

A basic radio circuit using a minimum of components is shown in Fig. 6.1. L1 and C1 is a conventional tuned circuit, e.g. a high-Q proprietary coil on a ferrite rod with a matching value of tuning capacitor. Alternatively, L1 can be made by winding approximately 80 turns of 0.3 mm diameter (30 swg) enamelled copper wire on a ferrite rod 4 cm ($1\frac{1}{2}$in) to 7.5 cm (3 in) long. In this case a matching value of C1 is 150 pF.

This circuit will provide sufficient output power for driving a

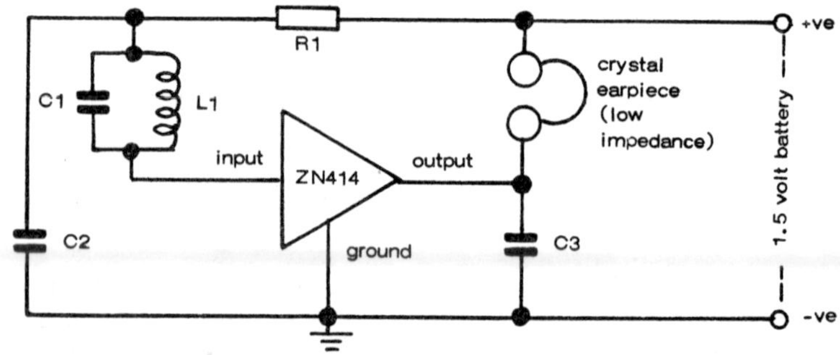

6.1 The ZN414 integrated circuit consists of a preamplifier followed by three stages of *rf* amplification and finally a transistor detector. It is a 'complete' radio circuit requiring a minimum of external components to work. These are:

L1 and C1 – ferrite rod aerial coil and matching tuning capacitor
R1 – 100 kohms
C2 – 0.01 μF
C3 – 0.1 μF

Sensitive (low impedance) crystal earpiece (500 ohms) or less.

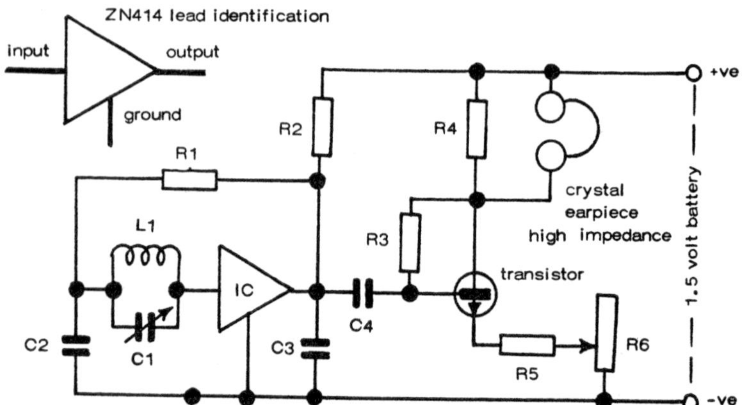

6.2 To deliver enough power to work a high impedance crystal earpiece the ZN414 is used in conjunction with an additional stage of transistor amplification. It delivers 600 mW peak output. This is the same circuit as Fig. 1.4 (Chapter 1) with the addition of a volume control (R6)

Component values:

R1 – 100 kohms
R2 – 1 kohm
R3 – 100 kohms
R4 – 10 kohms
R5 – 100 ohms
R6 – 250ohms potentiometer

L1 & C1 as Fig. 6.1
C2 – 0.01 μF
C3 – 0.1 μF
C4 – 0.1μF

sensitive low impedance earpiece with an equivalent resistance of approximately 500 ohms. To work a high impedance crystal earpiece an additional stage of amplification is needed. This modified circuit is shown in Fig. 6.2, requiring four more resistors, a potentiometer, another capacitor and a ZTX300 transistor (or equivalent). The potentiometer R4 and resistor R5 provide volume control (by adjustment of R4). This can be omitted if the receiver is to be brought down to minimum size, as the directional effects of the ferrite rod aerial will normally provide all the volume control necessary. In that case, replace R4 and R5 with a single 270-ohm resistor.

Fig. 6.3 shows the circuit extended to give a performance comparable to that of most domestic portable transistor receivers, driving an 8-ohm loudspeaker and formed by a 9-volt battery. This circuit does use six additional transistors and a number of other components, but the component count (and cost) is still substantially less than that of an all-transistor receiver of comparable quality (it is the equivalent of a 16 transistor set).

AM/FM Radio

A design for a high performance AM/FM radio receiver is shown in Figs. 6.4 and 6.5. These circuits are by Mullard and are based on their TDA 1071 integrated circuit which incorporates an AM oscillator, an AM mixer with *agc,* a four-stage differential amplifier and limiter and a four-quadrant multiplier. Both AM and FM functions are combined in the multiplier, giving symmetrical demodulation on AM and quadrative detection with squelch on FM.

Fig. 6.4 shows the AM circuit, working from a ferrite rod aerial. Fig. 6.5 shows the circuit for the additional front-end required for FM working, connected to an FM aerial. These circuits will work on any battery voltage from 4.5 volts to 9 volts. For FM operation, the AM-FM switch (SW4) moved to the FM position switches off the AM mixer and oscillator and brings the FM front-end circuit into operation. The squelch circuit is separately controlled by SW1, the threshold of squelch operation being set by the potentiometer R11 in Fig. 6.4.

Component values are given on the two circuit diagrams. A complete list is also given on pages 70 and 71.

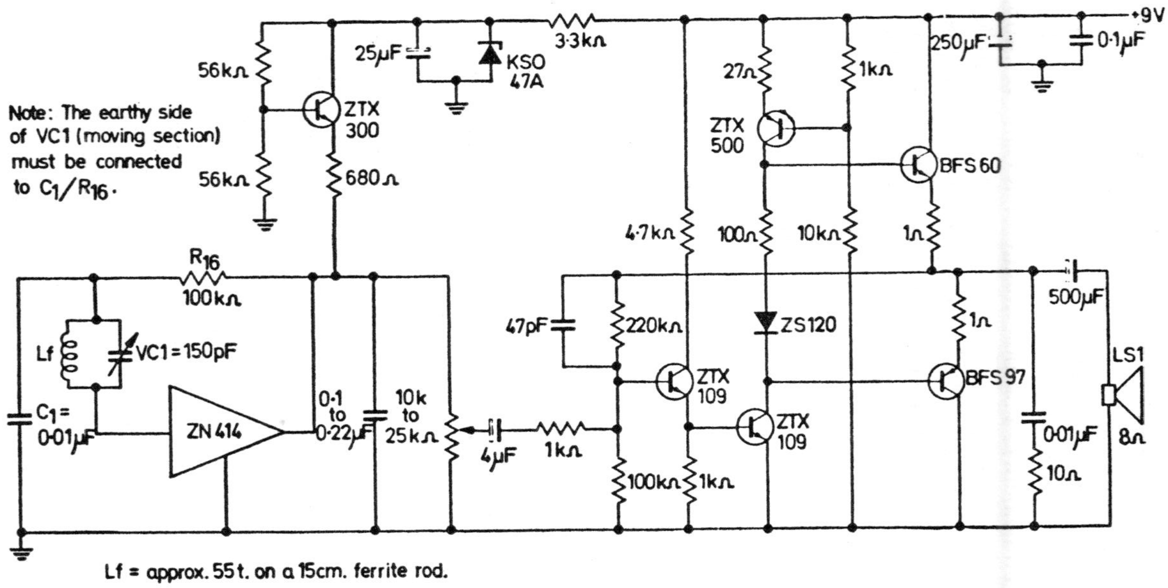

6.3 A high quality receiver circuit based on the ZN414 integrated circuit. This is a design by Ferranti. Component values are given on the circuit diagram. A 9-volt battery is used for the supply voltage.

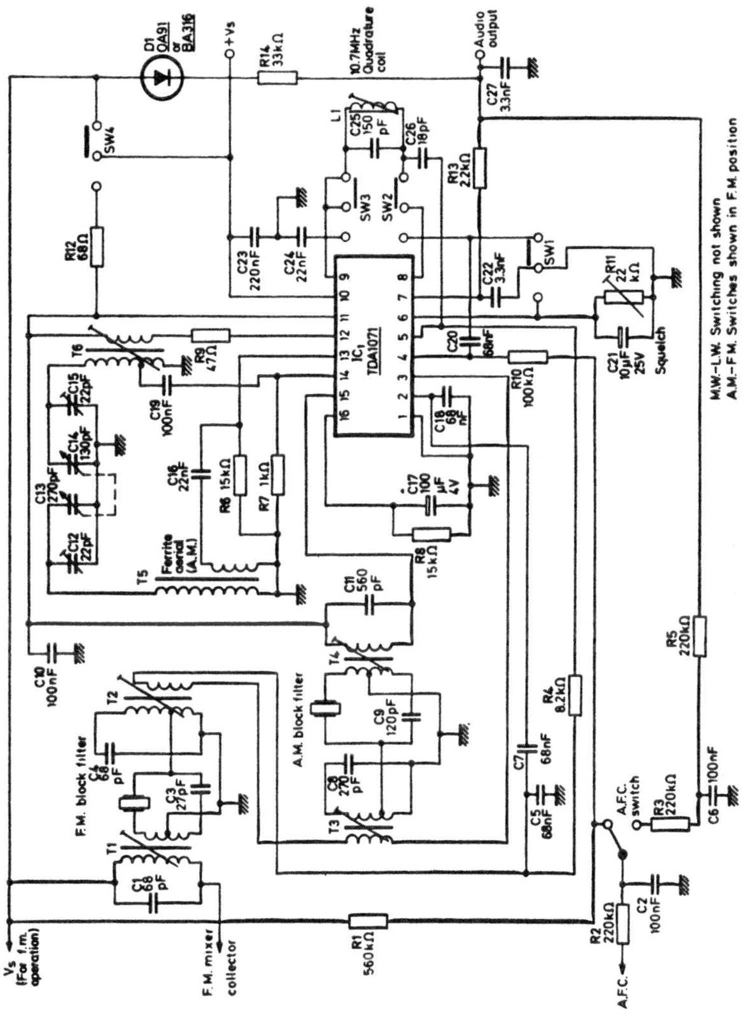

6.4 Circuit diagram of AM/FM receiver using the TDA1071 integrated circuit (Mullard)

6.5 Front-end circuit for FM operation of the receiver given in Fig. 6.4 (Mullard)

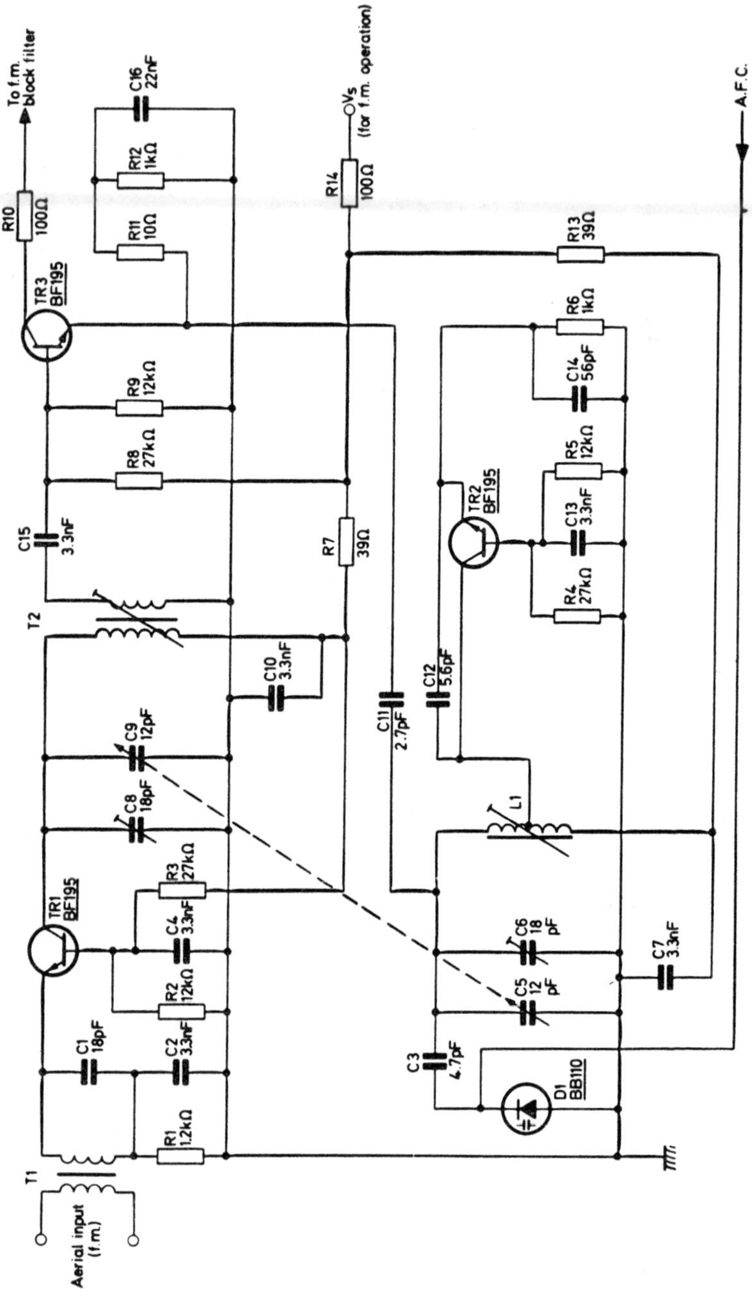

To f.m. block filter
R10 100Ω
TR3 BF195
C16 22nF
R12 1kΩ
R11 10Ω
R9 12kΩ
R8 27kΩ
C15 3.3nF
T2
Vs (for f.m. operation)
R14 100Ω
R7 39Ω
C10 3.3nF
C9 12pF
C8 18pF
TR1 BF195
R3 27kΩ
C4 3.3nF
R2 12kΩ
C2 3.3nF
C1 18pF
R1 1.2kΩ
T1
Aerial input (f.m.)
C11 2.7pF
C12 5.6pF
TR2 BF195
R13 39Ω
R6 1kΩ
C14 56pF
R5 12kΩ
C13 3.3nF
R4 27kΩ
L1
C6 18 pF
C5 12 pF
C7 3.3nF
C3 4.7pF
D1 BB110
A.F.C.

Components: AM/FM receiver circuit

Resistors

All resistors CR25 10% unless stated

R1	560kΩ
R2	220kΩ
R3	220kΩ
R4	8.2kΩ
R5	220kΩ
R6	15kΩ
R7	1kΩ
R8	15kΩ
R9	47Ω
R10	100kΩ
R11	22kΩ Miniature carbon preset potentiometer, Philips 2322 410 03309
R12	68Ω
R13	2.2Ω
R14	33kΩ

Capacitors

C1	68pF
C2	100nF
C3	27pF
C4	68pF
N5	68nF
C6	100nF
C7	68nF
C8	270pF
C9	120pF
C10	100nF
C11	560pF
C12	22pF
C13	270pF*
C14	130pF*
C15	22pF
C16	22nF
C17	100μF, 4V
C18	68nF
C19	100nF
C20	68nF
C21	10μF, 25V
C22	3.3nF
C23	230nF
C24	22nF
C25	150pF
C26	18pF
C27	3.3pF

*These components form part of the ganged tuning capacitor

Winding data

T1 Primary: 12 turns, 0.071mm enamelled copper
Secondary: 2 turns, tapped at 1 turn, 0.071mm enamelled copper
Former: Toko 7P 0092

T2 Primary: 12 turns, tapped at 1 turn, 0.071mm enamelled copper
Secondary: 3 turns, 0.071mm enamelled copper
Former: Toko 7P 0092

T3 Primary: 3 turns, 0.071mm enamelled copper
Secondary: 120 turns, tapped at 5 turns, wound over primary, 0.071mm enamelled copper
Former: Toko 7P 0089

T4 Primary: 9 turns, tapped at 5 turns, 0.071mm enamelled copper
Secondary: 86 turns, wound over primary, 0.071mm enamelled copper
Former: Toko 0089

T5: M.W. –
Primary: 78 turns, wound in a single layer, 3 x 3 x 3 x 0.063mm litz
Secondary: 4 turns, wound over the earthy end of the primary 3 x 3 x 3 x 0.063mm litz
L.W. –
Primary: 210 turns, wave-wound, 9 x 0.063mm litz
Secondary: 12 turns, wound under the primary, 9 x 0.063mm litz

For T5 the coils are mounted on a ferrite rod, 178mm in length, diameter 9.5mm.

L1 8 turns, 0.071mm enamelled copper. Former: Toko 7P 0092

Switch

SW1 to SW4 4-pole 2-way switch

Integrated circuit

IC1 TDA1071

Components: FM front-end circuit

Resistors
All resistors CR25 10%
R1 1.2kΩ
R2 12kΩ
R3 27kΩ
R4 27kΩ
R5 12kΩ
R6 1kΩ
R7 39Ω
R8 27kΩ
R9 12kΩ
R10 100Ω
R11 10Ω
R 1-
2 1kΩ
R 1-
3 39Ω

Winding data
T1 Primary: 2 turns, 0.031 mm enamelled copper
Secondary: 2 turns, 0.031 mm enamelled copper
Former: Neosid 5 mm with ferrite core
T2 Primary: 4 turns, spaced one diameter 0.71 mm enamelled copper
Secondary: 1 turn, interwound with the primary 0.71 mm enamelled copper
Former: Neosid 5 mm with ferrite core
L1 3 turns, spaced one diameter and tapped at 1½ turns, 0.71 mm enamelled copper
Former: Neosid 5 mm with ferrite core

Capacitors
C1 18pF
C2 3.3nF
C3 4.7pF
C4 3.3nF
C5 12pF*
C6 18pF
C7 3.3nF
C8 18pF
C9 12pF*
C10 3.3nF
C11 2.7pF
C12 5.6pF
C13 3.3nF
C14 56pF
C15 3.3nF
C16 22nF
*These components form part of the ganged tuning capacitor

Transistors
TR1, TR2, TR3 BF195

Diode
D1 BB110

Printed Circuit Layout

Fig. 6.6 shows a printed circuit layout for the complete circuits of Figs. 6.4 and 6.5, using the components specified above. Components with the subscript F are those in the front end circuit (Fig. 6.5). One additional component is also shown—a 300pF capacitor adjacent to the medium wave/long wave AM aerial switch, which does not appear on the relevant circuit diagram (Fig. 6.4).

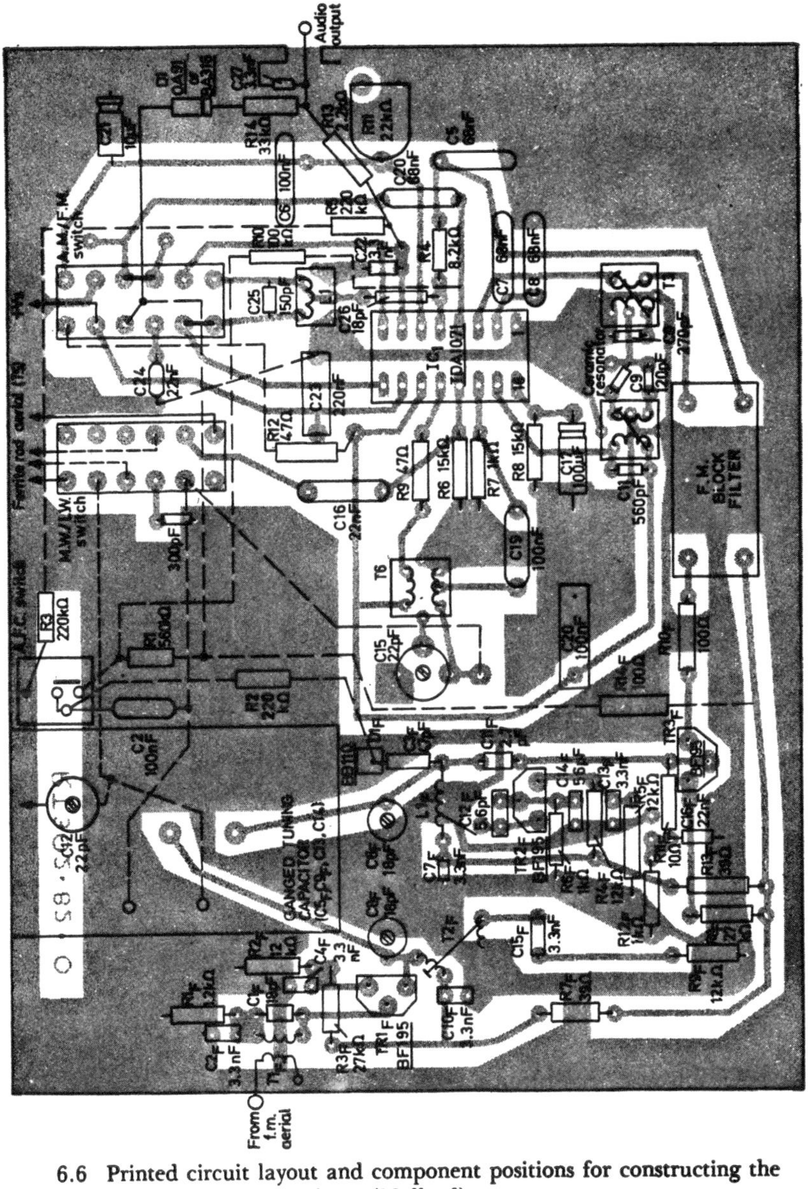

6.6 Printed circuit layout and component positions for constructing the circuits of Fig. 6.4 and 6.5. (Mullard)

Note also that this circuit is complete only up to the audio output stage—i.e. it needs to be followed by an audio amplifier and speaker(s)—*see* Chapter 4 for possible circuits to use.

Chapter Seven

MULTIVIBRATORS

THE SIMPLEST FORM of IC multivibrator merely uses an op-amp in a basic oscillator circuit such as that shown in Fig. 7.1.

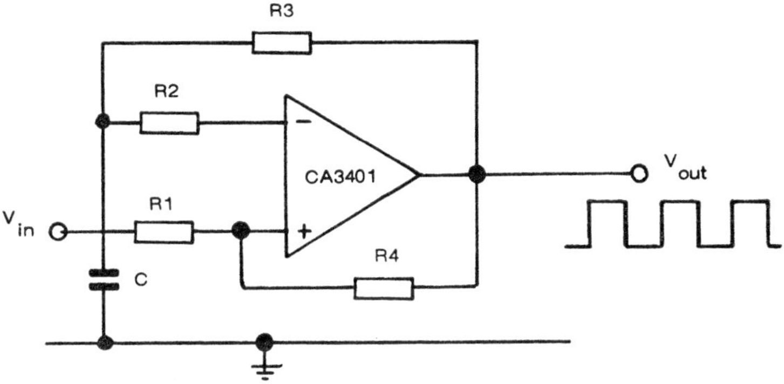

7.1 Simple multivibrator (or square wave oscillator) circuit based on the CK3401 op-amp. Component values may be chosen to give any specific output frequency required, within limits. The following component values give a 1 kHz square wave output.

R1 — 10 M ohm
R2 — 3 M ohm
R3 — 30 kohms
R4 — 10 M ohm
C — 0.01 μF

Oscillation frequency will depend on the IC parameter and the values of the external resistors. The components shown, the output frequency will be 1 kHz and in the form of a square wave.

The addition of a diode to this circuit, as in Fig. 7.2, provides a simple *pulse generator* circuit where the pulse width can be adjusted by using different values for R2. The value of resistor R3 governs the actual pulse duration.

An alternative form of multivibrator is to use two op-amps connected as cross-coupled inverting amplifiers, as shown in Fig. 7.3. Here the frequency is established by the time constants of the RC combination R1-C1 and R2-C2. R1 and R2 should be the same value, and can be anything from 1k ohm to 10k

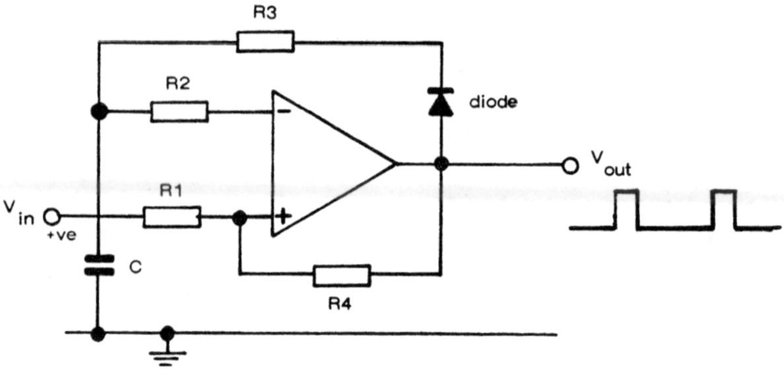

7.2 An almost identical circuit, with the addition of a diode, can be used as a pulse generator. Here the value of R3 determines the pulse duration and the value of R2 determines the 'off' period.

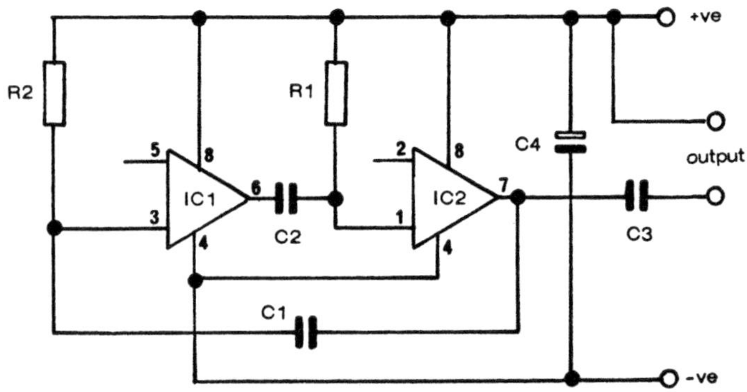

7.3 Multivibrator circuit based on the μL914 integrated circuit which is basically two inverting op-amps. The frequency of oscillation is determined by the time constants of R1, C2 and R2, C1. Suggested component values for working as an audio tone generator are:

R1 and R2 – 1 kohm to 10 kohm
C1 and C2 – 0.02μF to 2μF
C3 – 0.01μF
C4 – 100μF/12 volts
IC1 and IC2 – Fairchild μL914 (pins 2 and 5 not used)

Supply voltage 3.6 to 6 volts.

ohms. C1 and C2 should also be similar values, and anything from 0.01 to 10μF can be used. The basic rules governing adjustment and oscillation frequency are that for any particular value of R1 and R2, *increasing* the value of C1 and C2 will *decrease* the oscillation frequency, and vice versa. Similarly, for any particular value of C1 and C2 *decreasing* the value of R1 and R2 will *increase* the frequency, and vice versa.

With the component values shown, i.e. R1 = R2 = 8.2k ohms and C1 = C2 = 0.2μF, the oscillation frequency will be 1 kHz. Decreasing the value of R1 and R2 to 1k ohm should result in an oscillation frequency of 10 kHz.

A rather more versatile multivibrator circuit is shown in Fig. 7.4, which has independent controls of 'on' and 'off' periods.

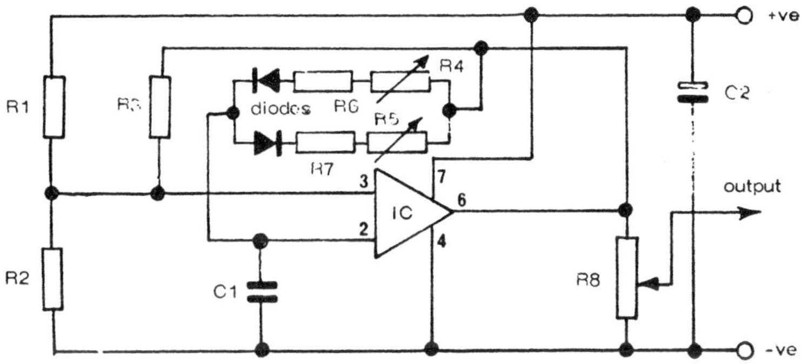

7.4 Multivibrator circuit with adjustable 'on' and 'off' periods.
Component values:

R1 – 100 kohms	C1 – *see* text
R2 – 100 kohms	C2 – 0.01 μF
R3 – 100 kohms	
R4 – 1 Mohm potentiometer	
R5 – 1 Mohm potentiometer	
R6 – 2 kohms	
R7 – 2 kohms	
R8 – 2 kohms potentiometer	
IC – CA3130	
Supply voltage 15 volts	

The frequency range is adjustable by choice of capacitor C1 which governs the duration of the square wave pulse generated, viz:

Value of C1	pulse period	frequency
1μF	4 min to 1 sec	250 - 1 Hz
0.1μF	0.4 min to 100 min	2500 - 600 Hz
0.01μF	4 min to 10 min	1500 - 6000 Hz
0.00μF	4 sec to 1 min	15000 kHz - 60 kHz

Adjustment of 'on' and 'off' times of oscillation within these ranges is governed by the potentiometers R4 and R5.

Another multivibrator circuit is shown in Fig. 7.5, which is particularly notable for its stable performance. The frequency of oscillation is maintained to within plus or minus 2 per cent

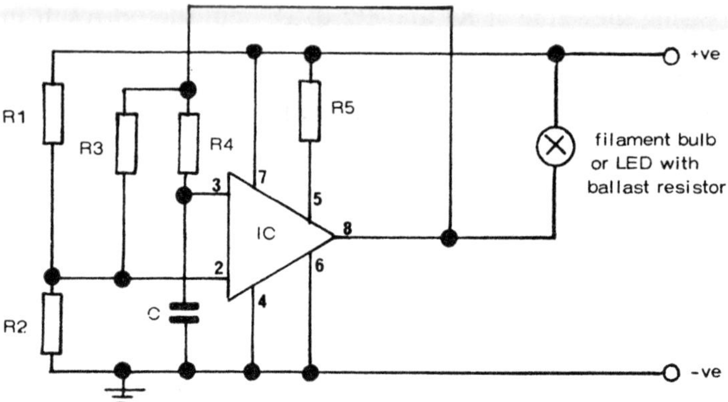

7.5 Astable multivibrator with low frequency of duration to work a flashing light. Component values given provide a flashing rate of about 1 per second working off a 6 to 15 volt battery.

R1 — 3 M ohms
R2 — 12 M ohms
R3 — 18 M ohms
R4 — 4.3 M ohms
R5 — 1.2 M ohms
C — 0.45 μF
IC — CA3094
Filament bulb — 6 or 12 volts, depending on battery voltage.

on any supply voltage from 6 to 15 volts and is independent of the actual voltage. It uses a CA3094 op-amp IC with external resistors and one capacitor. The circuit also includes a lamp which flashes on and off at a rate of one flash per second with the component values given.

Flashing rate can be adjusted by altering the values of R1 and R2 and/or C. To adjust values to give any required flashing rate (frequency), the following formula applies:

$$\text{frequency} = \frac{1}{2RCI_n(2R1/R2 + 1)}$$

$$\text{where } R1 = \frac{RA \, . \, RB}{RA + RB}$$

In a variation on this circuit shown in Fig. 7.6, the introduction of a potentiometer R2 enables the pulse length to be

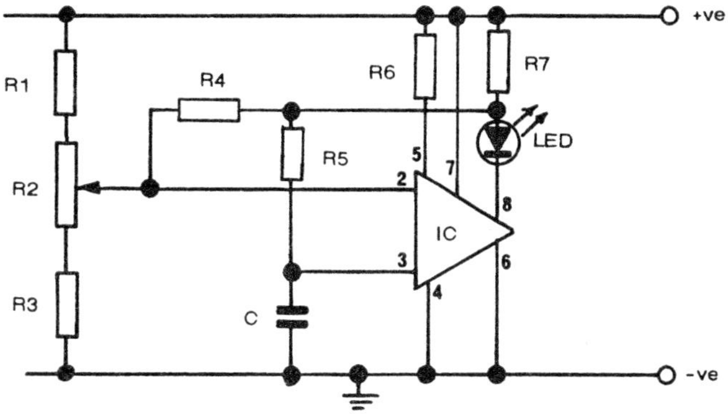

7.6 Adjustable multivibrator circuit, potentiometer R2 varying the pulse width, or 'on' time of the LED indicator.
Flashing rate is approximately 1 per second. Supply voltage required for this circuit is 22 to 30 volts.
Component values:
R1 — 27 kohms
R2 — 50 kohms potentiometer C — 560 pF
R3 — 27 kohms
R4 — 100 kohms
R5 — 100 kohms
R6 — 300 kohms
R7 — ballast resistor to suit LED used
IC — CA309A
LED — light emitting diode

varied whilst maintaining a constant frequency (pulse repetition rate). Again this circuit can be used to flash a filament lamp, or a light emitting diode. In the latter case, a ballast resistor is needed in series with the LED.

Another straightforward free-running multivibrator is shown in Fig. 7.7, using a CA3094 integrated circuit. The frequency is controlled by the value selected for R3, and so using a potentiometer for this component enables the frequency to be adjusted. The frequency is also dependent on the supply voltage, which can be anything from 3 volts up to 12 volts.

Designing a multivibrator circuit to work at an audio frequency, whilst retaining adjustment of frequency, forms the basis of a *metronome.* The only additional circuitry required is a simple low-power audio amplifier connecting to a loud-speaker of the kind described in Chapter 4.

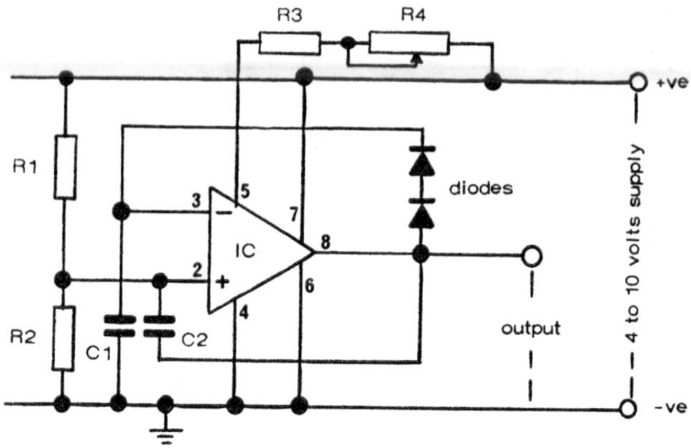

7.7 Free running multivibrator (or pulse generator) circuit, the frequency of which can be varied by adjustment of the potentiometer R4.
Component values:
R1 – 2.7 M ohms
R2 – 330 k ohms
R3 – 100 k ohms
R4 – 5 k or 10 k ohms potentiometer
C – 0.01 μF
diodes – any silicon diodes
IC – CA3094

Chapter Eight

VOLTAGE REGULATORS

WHERE A SEMICONDUCTOR CIRCUIT operating on low voltage *dc* is powered from the mains supply via a step-down transformer, voltage regulation is highly desirable in many circuits in order to ensure constant *dc* supply voltage. This can be provided by using Zener diodes in associated circuitry. Exactly the same function can be performed by a single IC. A particular advantage is that such an IC can also incorporate internal overload and short-circuit protection which would call for numerous extra components in a circuit using discrete components.

A typical circuit is shown in Fig. 8.1, giving a rectified, positive *dc* voltage output from the centre tapped secondary of

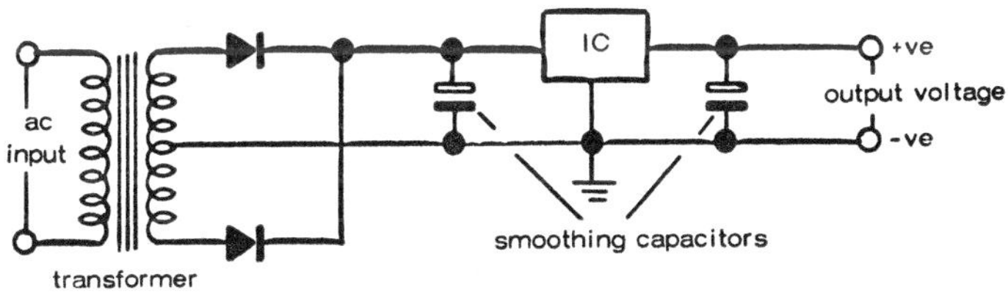

8.1 Typical basic circuit for stepping down an *ac* voltage via a transformer and rectifying it to produce a lower voltage *dc* output. A voltage regulating IC is also shown in this circuit. Note that a rectangular symbol is used for the IC in this case, not a triangle.

the transformer. The same components can be used in mirror-image configuration to give a negative output voltage from the centre tap (in which case the polarity of the two electrolytic capacitors must be reversed).

Performance characteristics of a family of ICs designed as voltage regulators are given on p. 80. They are quite small devices in a TO-39 metal case with three leads—input, output and earth—*see* Fig. 8.2. The earth or ground lead is internally connected to the case.

IC type no.	input voltage	output voltage	Max. output current
TBA 435	20	8.5	200 mA
TBA 625A	20	5	200 mA
TBA 625B	27	12	200 mA
TBA 625C	27	15	200 mA

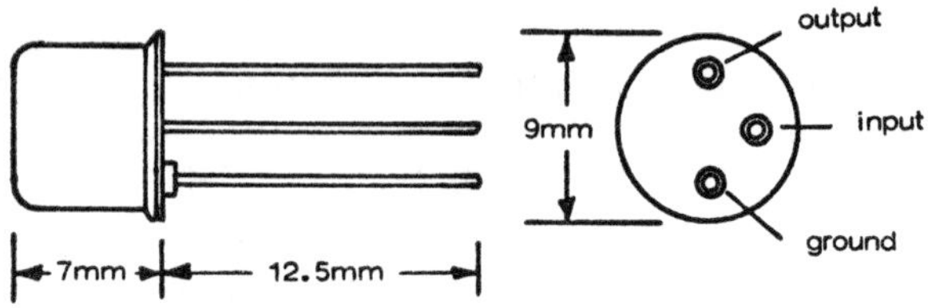

8.2 TBA435 integrated circuit is enclosed in a TO-39 metal can shape and looks like a transistor because it only has three leads. It is a complete voltage regulator circuit with internal overload and short circuit protection and can be used in the circuit of Fig. 8.1. The drawing is approximately twice actual size.

There are numerous other simple voltage regulators which can be built from integrated circuit arrays (*see* Chapter 2) simply by 'tapping' the appropriate leads to connect the individual components required into the complete circuit. An example is shown in Fig. 8.4, which is a regulator to provide an adjustable constant voltage *dc* output from an unregulated (and thus possibly variable) 20 volt *dc* input. It uses the transistors, diode and Zener diode contained in the CA 3097E chip with a potentiometer and external resistor to complete the circuit. The actual output voltage can be adjusted from 9.5 to 15 volts by the setting of the potentiometer, with an output current ranging up to 40 mA, depending on the value of the output load.

Other simple voltage regulators can be based on op-amps. A basic circuit is shown in Fig. 8.5. The reference voltage is set by the Zener diode, the value of R1 being chosen to provide optimum Zener current for the input voltage concerned. The (regulated) output voltage is determined by the reference voltage (V ref) and the values of R2 and R3: —

$$V\ out = V\ ref\left(\frac{R2 + R3}{R3}\right)$$

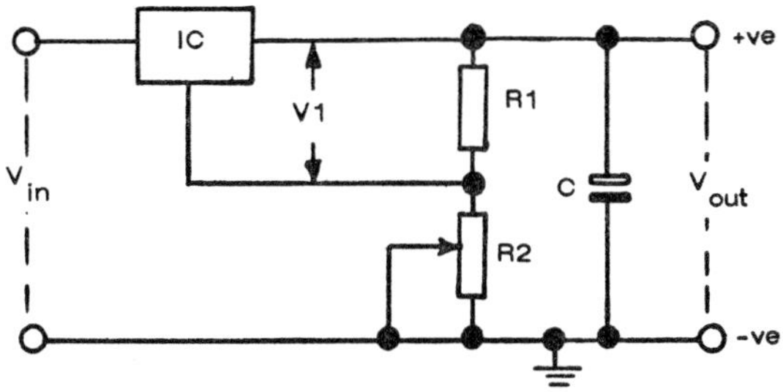

8.3 Adjustable output voltage regulator circuit. The potential divider formed by R1 and R2 following the IC enables the output voltage to be adjusted via R2; otherwise the circuit is the same as Fig. 8.1. Alternatively this circuit can be applied to a *dc* input:

$$V\,out = V1\,(1 + \frac{R2}{R1}) + 1_G R2$$

Component values for this circuit with an input voltage of 18 volts:

R1 — 680 ohms
R2 — 0-150 ohms potentiometer
C — 10μF
IC — TBA435

Note. Other IC voltage regulators can be used and/or different input voltages, in which case different values of R1 and R2 may apply. As a general rule R2 needs to be about one-third to one-half the value of R1.

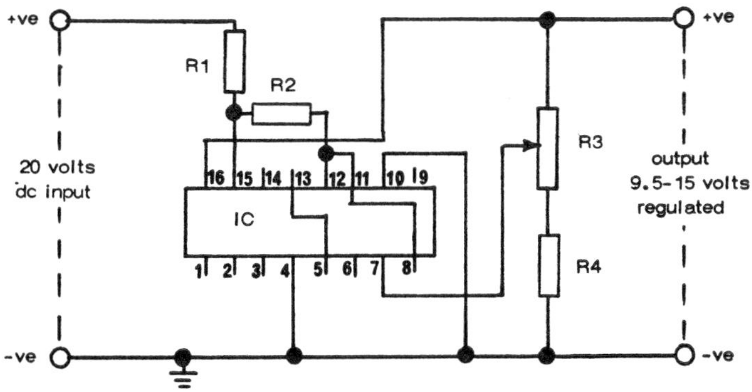

8.4 Voltage regulator using the CA3097E integrated circuit array. This provides 9.5 to 15 volts regulated output from a 20 volts *dc* input, the actual output voltage being determined by the setting of R3.

Component values:

R1 — 2 kohms
R2 — 2 kohms
R3 — 2.5 kohms potentiometer
R4 — 1.5 kohms

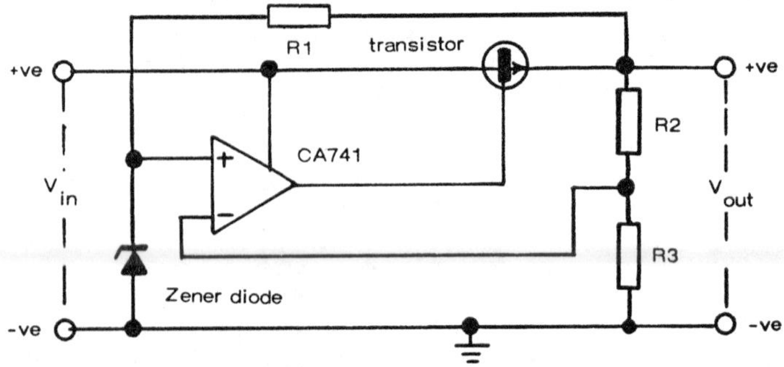

8.5 Simple voltage regulator circuit using an op-amp and a Zener diode to set the regulated voltage. The values of resistors R2 and R3 determine the output voltage (*see* text).

A circuit which provides a small difference between *volts in* and *volts out* is shown in Fig. 8.6. Using a PNP transistor it needs only about 1 volt to saturate the transistor, whilst adequate current is available for the regulating circuit using an NPN transistor. The same circuit would need about 4 volts difference between input and output to maintain regulation.

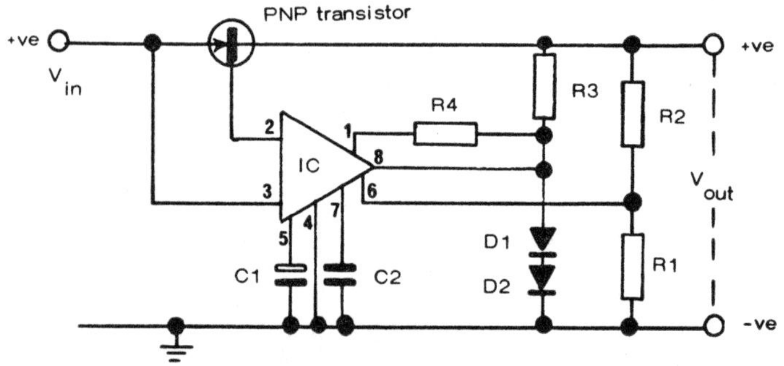

8.6 Voltage regulator circuit incorporating a PNP transistor which gives a difference between V in and V out of about 1 volt (i.e. the voltage necessary to saturate the transistor).

Component values:

R1 – } *see* text
R2 – }
R3 – 10 kohms
R4 – 5 kohms potentiometer
IC – CA3085
transistor – 40362 (or equivalent)
D1, D2 – silicon diodes

C1 – 2μF
C2 – 0.01μF

With the circuit shown in Fig. 8.6,

$$\text{volts out} = 1.6\,\frac{(R1 + R2)}{R1}$$

Another very useful circuit is shown in Fig. 8.7, which provides a split supply from a single battery source. In other words it halves the input voltage whilst also producing a good degree of regulation of the two (plus and minus) voltage outputs. None of the component values is critical but R1 and R2 should be of close-tolerance type of equal value. Input voltage can range from 6 to 36 volts, when one half of the input voltage will appear between output + and 0, and the other between 0 and output –.

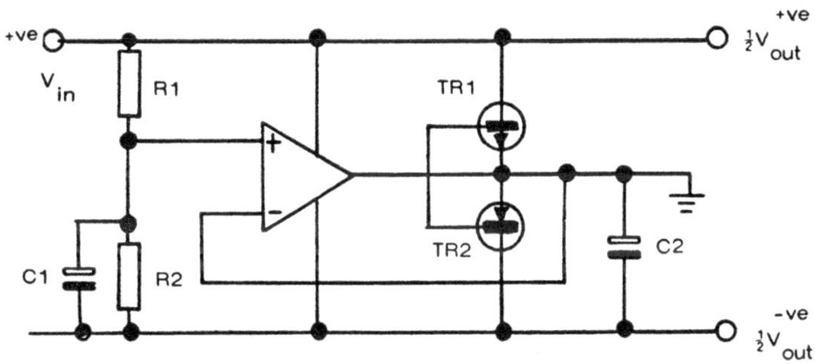

8.7 This circuit gives a regulated split supply from any input voltage from 5 to 36 volts.

Component values:

R1 – 180 kohms (close tolerance)
R2 – 180 kohms (close tolerance)
C1 – 25μF
C2 – 25μF
TR1 – ZN1711 (or equivalent)
TR2 – 40362 (or equivalent)
IC – CA741

Chapter Nine

ELECTRIC MOTOR SPEED CONTROLLERS

A VARIETY OF ICS are designed as speed regulators for small *dc* motors such as those used in portable cassette players, movie cameras, models and toys. The object is to 'govern' the motor so that it runs at a constant speed, independent of variations in battery supply voltage and load on the motor. The TDA1151 is selected for the following circuits, having a maximum rating of 20 volts (which covers most model and other small *dc* motors), with an output current of up to 800 milliamps. It is a flat rectangular plastic package with three leads emerging from one end, and comprises 18 transistors, 4 diodes and 7 resistors in a linear integrated circuit.

In its simplest application it is used with a potentiometer (R_s) acting as a speed regulation resistance (and by which the actual motor speed is adjusted); and a torque control resistor (R_t) which provides automatic regulation against load on the motor. Both these resistors are bridged by capacitors, although C2 can be omitted—*see* Fig. 9.1. Component values shown are suitable for a 6 to 12 volt supply.

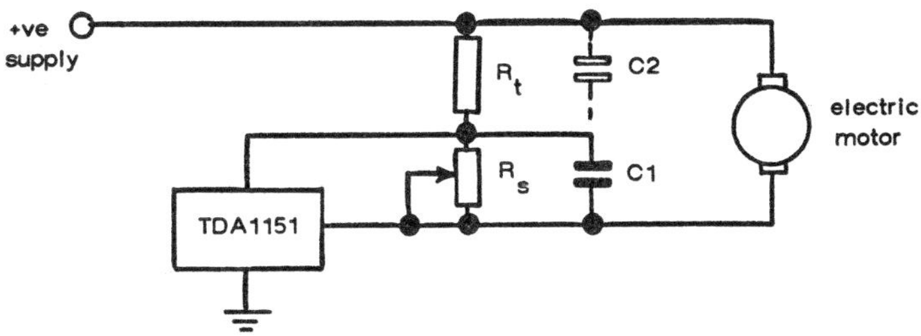

9.1 Use of the TDA1151 linear integrated circuit as a speed regulator for a small *dc* electric motor.
Typical component values:
R_s – 1 kohm
R_t – 280 ohms
C1 – 10 mF to 2 μF
C2 (if used) – 25 μF

A slightly different circuit is shown in Fig. 9.2, using a TCA600/900 or TCA610/910 integrated circuit. These have maximum voltage ratings of 14 and 20 volts respectively; and maximum current ratings of 400 milliamps for starting, but only 140 milliamps for continuous running.

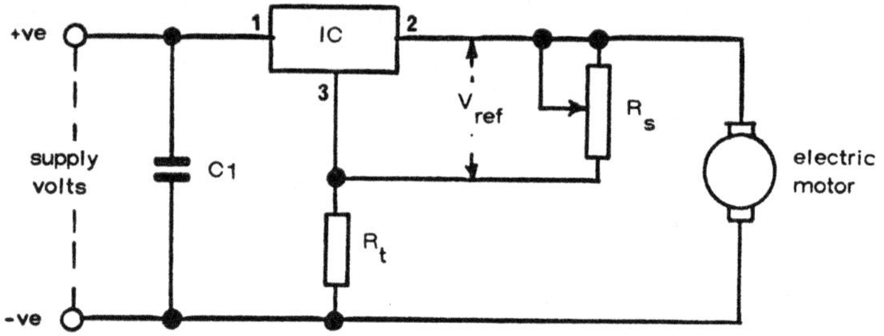

9.2 Application circuit for the TCA600/610 or TCA900/910 motor speed regulators. R_s is the speed regulation resistor (variable). R_t is the torque control resistor. A suitable value for Cl is 0.1 F. A diode can be added in line 3 to provide temperature compensation as well.

Devices of this type work on the principle of providing a constant output voltage to the motor independent of variations of supply voltage, the value of this voltage being set by adjustment of R_s. At the same time the device can generate a negative output resistance to compensate speed fluctuations due to variations in torque. This negative output resistance is equal to RT/K, where K is a constant, depending on the parameters of the device, viz:

IC	K (typical)	Vref	I_o
TDA1151	20	1.2	1.7 mA
TCA600/900	8.5	2.6	2.6 mA
TCA610/910	8.5	2.6	2.6 mA

The above also shows the reference voltage (Vref) and quiescent current drain (I_o) of the three ICs mentioned.

The following relationships then apply for calculating suitable component values for these circuits:

$$R_t = K.R_M$$

where R_M is the typical motor resistance

$$\text{minimum value of Rs} = \frac{Vref\,.\,RT}{Eg - (Vref - I_oRT)}$$

where Eg = back *emf* of motor at required or rated speed

I_o = quiescent current drain of the device

Actual voltage developed across the motor is given by:

$$\text{Volts (at motor)} = R_M . I_M + Eg$$

where I_M is the current drain by the motor at required or rated speed

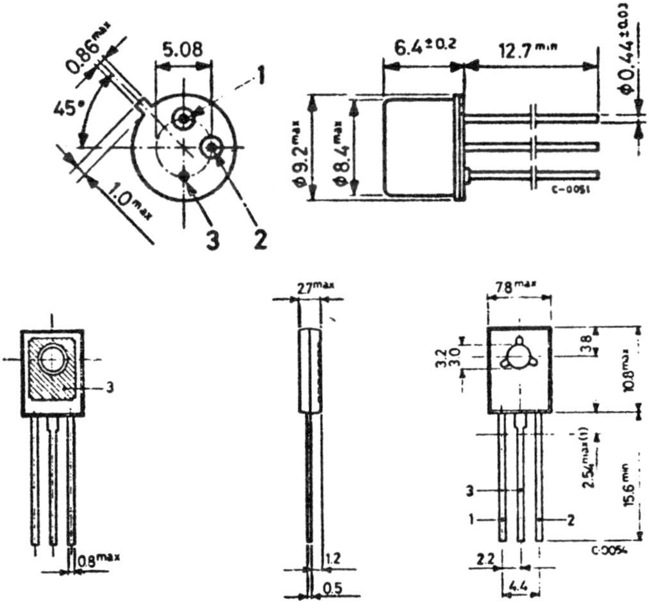

9.3 Physical appearance of the TCA600/610 in TO-39 metal can and TCA900/910 in flat plastic package (TO-126).

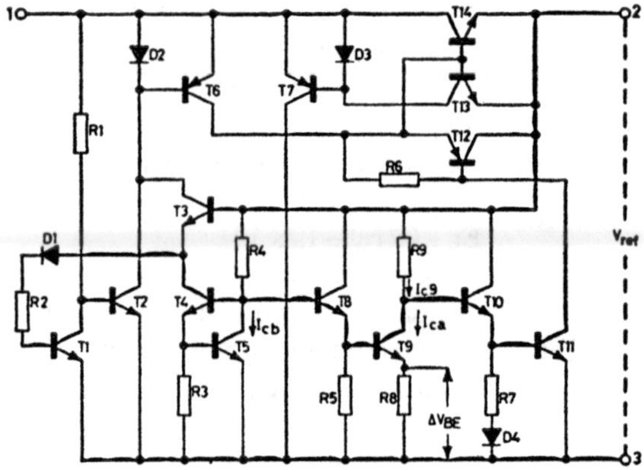

9.4 Although small devices, these integrated circuits for motor speed regulation are based on quite complicated circuitry. This diagram shows the internal circuits.

Chapter Ten

FILTERS

A BASIC FILTER CIRCUIT consists of a combination of a resistor and a capacitor. This combination of R and C has a *time constant* which defines the *cut-off frequency* of the filter; but the actual mode of working also depends on the configuration of the two components—*see* Fig. 10.1.

With R in series and C across the circuit, frequencies *lower*

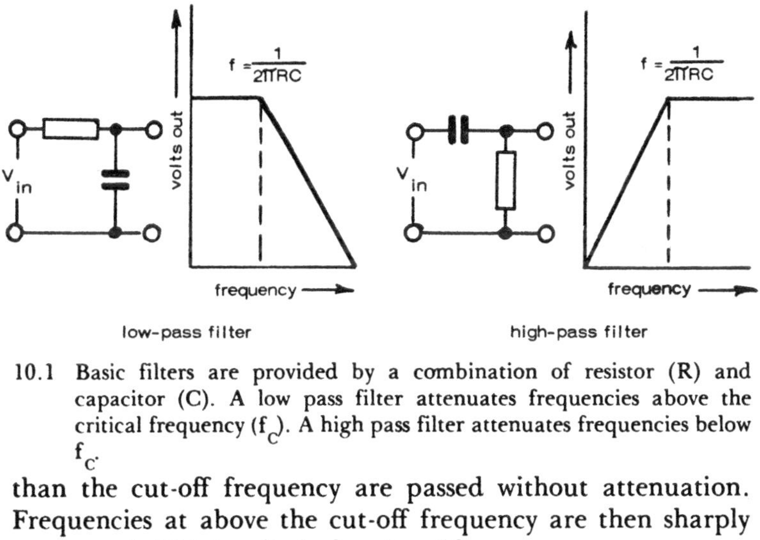

10.1 Basic filters are provided by a combination of resistor (R) and capacitor (C). A low pass filter attenuates frequencies above the critical frequency (f_C). A high pass filter attenuates frequencies below f_C.

than the cut-off frequency are passed without attenuation. Frequencies at above the cut-off frequency are then sharply attenuated. This is called a *low-pass filter*.

With C in series and R across the circuit, frequencies *above* the cut-off frequency are passed without attenuations. Frequencies below the cut-off frequency are then sharply attenuated. This is called a *high-pass filter*. Practical circuits for Practical circuits for these two types of filter are shown in Fig. 10.2.

The amount of attentuation provided by a filter is expressed by the ratio volts out/volts in, or voltage ratio. This is quoted in decibels (dB)—a 3 dB drop being equivalent to a voltage ratio drop from 1.0 to 0.707, or a *power* loss of 50 per cent.

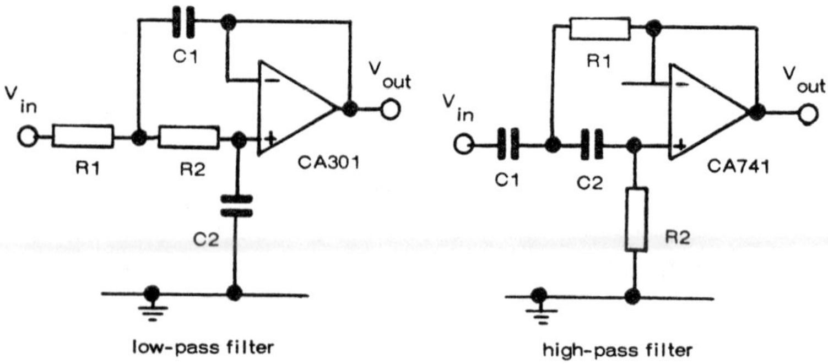

10.2 Basic low-pass and high-pass filter circuits incorporating an op-amp for better performance.

Op-amps can be used as practical filters associated with an external capacitor, with the advantage that the more sophisticated circuitry involved can provide superior performance to straightforward RC combinations.

Two filter circuits based on the CA301 op-amp are shown in Fig. 10.2. In the case of the low-pass filter component values are calculated from the formula:

$$C1 = \frac{R1 + R2}{1.414\, R1\, R2\, f_C}$$

$$C2 = \frac{1.414}{(R1 + R2)\, f_C}$$

where f_C is the effective cut-off point

In the case of the high-pass filter circuit:

$$C1 = \frac{R1 + R2}{1.732\, R1\, R2\, f_C}$$

$$C2 = \frac{1.732}{(R1 + R2)\, f_C}$$

Bandpass filters or bandwith filters can be produced by combining a low-pass filter in series with a high-pass filter. If the band width is from f_L to f_H, then the cut-off frequency for the low-pass filter is made f_H and that of the high-pass filter f_L—Fig. 10.3 (left). This filter combination will pass frequencies from f_L to f_H, i.e. in the desired band.

To produce a *band-reject filter,* a low-pass filter is used in parallel with ahigh-pass filter, as in the second diagram. This combination will reject all frequencies within the band f_L to f_H.

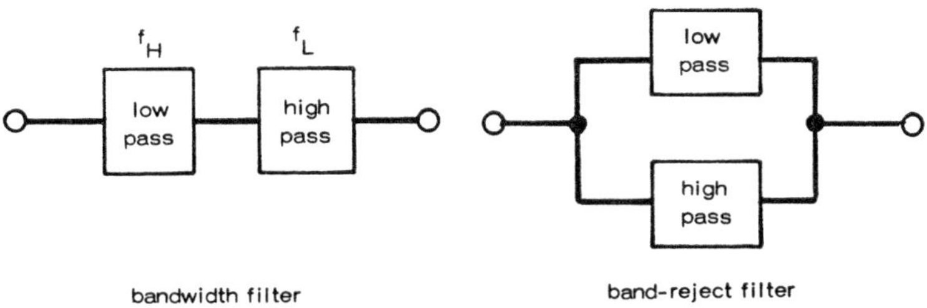

10.3 A low-pass filter in series with a high-pass filter passes frequencies only within the bandwidth $f_L - f_H$. A low pass filter in parallel with a high-pass filter rejects all frequencies with the band width $f_L - f_H$.

Some other practical filter circuits using op-amps are given in Figs. 10.4, and 10.5.

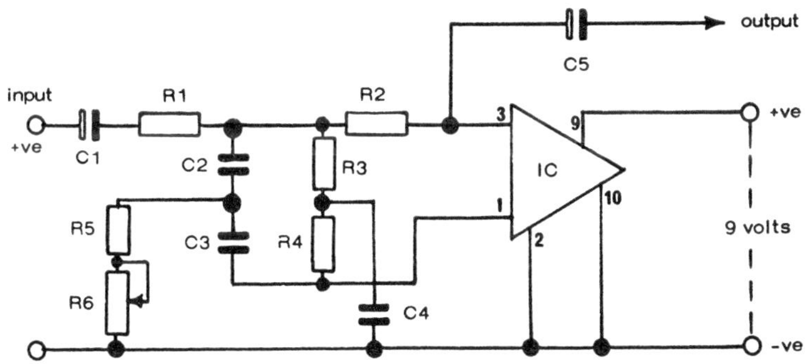

10.4 A notch filter rejects input signals at a specific centre frequency but passes all other frequencies, This is a working circuit, the centre frequency being determined by the value of components in the two networks R3-R4-R5-R6; and C2-C3-C4. The actual 'sharpness' of rejection or notch width is adjustable via potentiometer R6.
Component values for a 1 kHz centre frequency are:

R1 – 18 kohms	C1 – 10μF
R2 – 18 kohms	C2 – 0.001μF
R3 – 150 kohms	C3 – 0.001μF
R4 – 150 kohms	C4 – 0.001μF
R5 – 56 kohms	C5 – 10μF
R6 – 50 kohms potentiometer	
IC – CA3035	

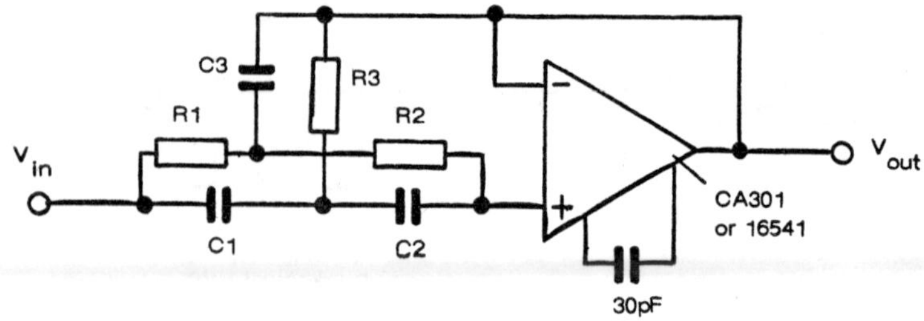

10.5 Simple circuit for a high Q notch filter. Capacitor C1 and C2 are equal in value. Capacitor C3 = C1/2. Resistor R1 is twice the value of R2. The centre frequency is

$$f_c = \frac{1}{2\pi R1\, C1}$$

Chapter Eleven

INTRODUCING DIGITAL CIRCUITS

THE DIGITAL SYSTEM (also known as the binary system) is based on counting in 1's. Thus it has only two digits (known as 'bits') — 0 (zero) and 1 (one) — which are very easy to manipulate electronically. It only needs a simple on-off switch, for example, to count in this manner. The switch is either 'off' (showing zero as far as the circuit is concerned) or 'on' (representing a count of 1). It can continue to count in 1's, or even multiply, divide, etc, in association with other simple types of switches. The fast-as-light speed at which electronic devices can count makes the digital system very suitable for building computer circuits, particularly as only a few basic operations have to be performed. The fact that these operations, using simple logic circuits or *gates* in suitable combinations, may have to be repeated very many times is no problem either.

The decimal system expresses a number in powers of 10. In other words individual digits, depending on their order represent the digit value $\times 10^0$, digit value $\times 10^1$, digit value $\times 10^2$, etc, reading from *right to left*. Putting this the correct way round, and taking an actual number—say 124:

$$\begin{aligned} 124 &= 1\times 10^2 + 2\times 10^1 + 4\times 10^0 \\ &= 100 + 20 + 4 \end{aligned}$$

The binary system expresses a number in powers of 2 using only the two digits 1 and 0.

$$\begin{aligned} \text{Thus } 1011 &= 1\times 2^3 + 0\times 2^2 + 0\times 2^1 + 1\times 2^0 \\ &= 8 + 0 + 2 + 1 \\ &= 11 \end{aligned}$$

Thus a binary number is longer, written down, than its corresponding decimal number, and can get very long indeed with large decimal numbers (e.g. 10,000 = 1010100010000) but this does not matter at all as far as electronics 'counting' is concerned. It only makes it difficult for people to convert decimal numbers to binary numbers, and vice versa. Here are two basic rules.

Converting decimal to binary

Write the decimal number on the right-hand side, divide by two and write down the result, placing the remainder (0 or 1) underneath this number. Divide the number obtained in the top line by 2 and carry the remainder (0 or 1) down to make a next step to the left. Repeat this operation, progressing further to the left each time, until you are left with an 0 in the top line.

Example

	0	1	2	4	9	19 decimal
this is the binary number	1	0	0	1	1	remainder

Converting binary to decimal

Write down progressively from *right to left* as many powers of 2 as there are digits in the binary number*. Write the binary number underneath. Determine the powers of 2 in each column where a 1 appears under the heading and then *add* all these up.

Example

Binary number 10101, which has 5 digits, so write down five stages of powers of 2 starting with 2^0 and reading from right to left.

	2^4	2^3	2^2	2^1	2^0
Write down binary number	1	0	1	0	1
Convert to decimal	16	0	4	0	1
Add	16+4+1=21				

Logic

Logic systems also work on the binary number process, commonly based on the difference between two *dc* voltage levels. If the more positive voltage signifies 1, then the system employs *positive* logic. If the more negative voltage signifies 1, then the system employs *negative* logic. It should be noted that in both cases, although the lower or higher voltage respectively signifies 0, this is not necessarily a *zero* voltage level, so the actual voltage values have no real significance.

There is another system, known as *pulse-logic*, where a 'bit' is

(*A group of binary digits or 'bits' which has a certain significance, i.e. represents a binary number in this case, is often called a 'bite' or 'word'.

recognized by the presence or absence of a pulse (positive pulse in the case of a positive-logic system and negative pulse in the case of a negative-logic system).

Gates

Logic functions are performed by logic *gates*. The three basic logic functions are OR, AND and NOT. All are designed to accept two or more *input* signals and have a single *output* lead. The presence of a signal is signalled by 1 and the absence of a signal by 0.

The four possible states of an OR gate with two inputs (A and B) are shown in Fig. 11.1. There is an *output* signal whenever

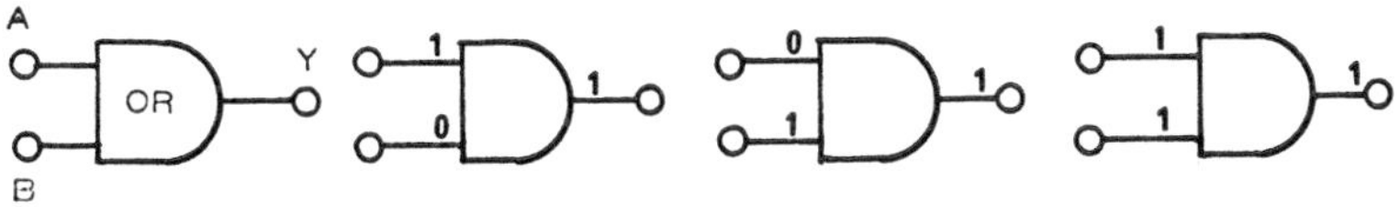

11.1 The three states of an OR gate. A and B are inputs and Y is the output. Note the general symbol used to illustrate a gate. For compactness a semicircle may be used instead of the symbol shown here (e.g. *see* Fig. 12.1).

there is an input signal applied to input A *OR* input B (and also with input at A and B simultaneously). This applies regardless of the actual number of inputs the gate is designed to accept. The behaviour of an OR gate (again written for only two inputs) is expressed by the following *truth table*:

A	B	output (Y)
0	0	0
0	1	1
1	0	1
1	1	1

It can also be expressed in terms of Boolean algebra, calling the output Y

$$Y = A + B + \ldots\ldots + N$$

where N is the number of gates

The important thing to remember is that in Boolean algebra the sign + does not mean 'plus' but OR.

The AND gate again has two or more inputs and one output, but this time the output is 1 only if *all* the inputs are also 1. The

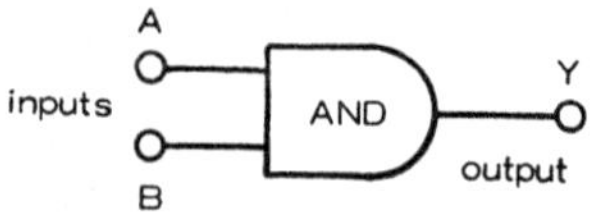

A	B	Y
0	0	0
1	0	0
0	1	0
1	1	1

TRUTH TABLE

11.2 An AND gate and corresponding truth table.

truth table in this case is quite different—Fig. 11.2. The corresponding equation of an AND gate is:

$$Y = A \cdot B \ldots\ldots N$$
$$\text{or } Y = A \times B \ldots\ldots xN$$

This time the · or × sign does not mean 'multiplied by' as in conventional arithmetic, but AND.

The NOT gate has a single input and a single output—Fig. 11.3, with output always opposite to the input, i.e. if A = 1, Y =

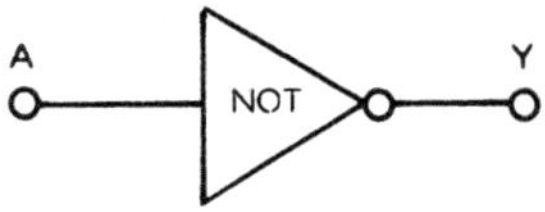

A	Y
0	1
1	0

TRUTH TABLE

11.3 A NOT gate and corresponding truth table. Note the symbol used in this case is the same as that for an op-amp or amplifier, and the following small circle designates an inverted output.

0 and if A = 0, Y = 1. In other words it inverts the sense of the output with respect to the input and is thus commonly called an *inverter*.
Its Boolean equation is:

$$Y = \overline{A}$$
(Y equals NOT A)

Combinations of a NOT gate with an OR gate or AND gate produce a NOR and NAND gate, respectively, working in the inverse sense to OR and AND.

Diode-logic (DL) circuits for an OR gate and an AND gate are shown in Fig. 11.4. Both are shown for negative logic and are identical except for the polarity of the diodes. In fact a positive-logic DL or OR gate becomes a negative-logic AND gate; and a positive-logic AND gate a negative-logic OR gate.

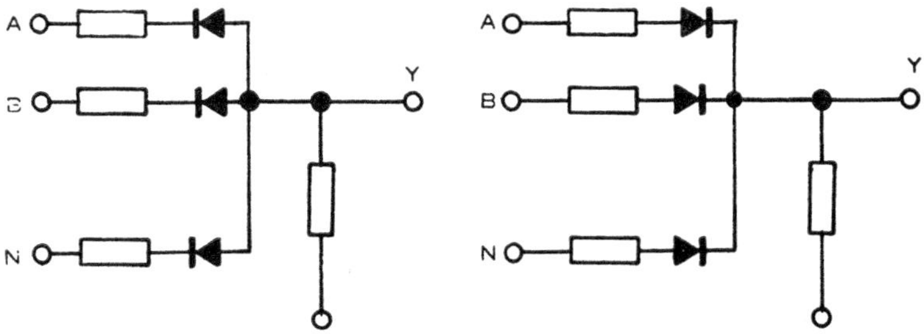

11.4 A Diode Logic (DL) negative logic OR circuit (*left*) and a DL negative logic AND gate (*right*).

The simple NOT gate or inverter shown in Fig. 11.5 is based on a transistor logic—an NPN transistor for positive-logic and a PNP transistor for negative-logic. The capacitor across the input resistance is added to improve the transient response.

Practical Gates

Most logic gates are produced in the form of integrated circuits, from which various 'family' names are derived. NAND

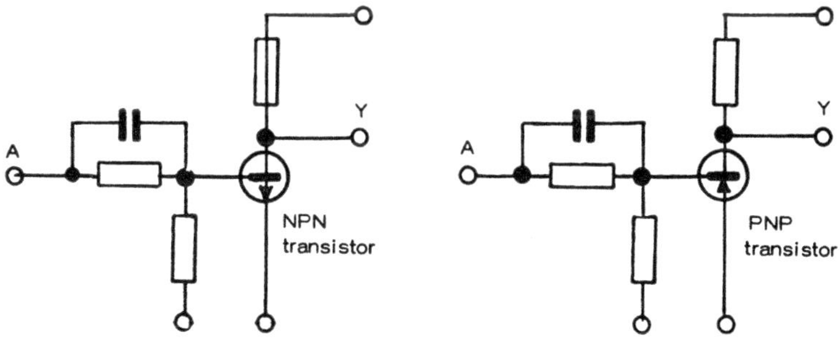

11.5 Transistor Logic (TL) positive logic inverter circuit (*left*) and a TL negative logic NOT circuit (*right*).

and NOR gates, for example, are a combination of AND or OR gates, respectively, with a NOT gate inverter. From the basic circuits just described, such functions can be performed by diode-transistor logic or DTL gates.

Faster and rather better performance can be realized with transistor-transistor-logic gates (TTL). During the early 1970's DTL and TTL represented the bulk of the IC digital productions, but since then various other IC families have appeared, each offering specific advantages and more functions for particular applications. These are:

RTL (resistor-transistor logic) which can be made very small—even by microelectronic standards—and is capable of performing a large number of functions.

DCTL (direct-coupled-transistor logic), which employs the same type of circuit as RTL but with the base resistors omitted. This gate, which can perform NOR or NAND functions, has the advantage of needing only one low voltage supply and has low-power classification.

HTL (high threshold logic) is based on diode-transistor logic similar to DTL but also incorporates a Zener diode to stabilize the circuit and provide high immunity to 'noise'. It is usually chosen for applications where this feature is important.

MOS (metal oxide semiconductor logic), based entirely on field effect transistors (FETs) to the complete exclusion of diodes, resistors and capacitors, yielding NAND and NOR gates.

CMOS (complementary metal-oxide-semiconductor logic) using complementary enhancement devices on the same IC chip, reducing the power dissipation to very low levels. The basic CMOS circuit is a NOT gate (inverter), but more complicated NAND and NOR gates and also flip-flops can be formed from combinations of smaller circuits (again in a single chip).

ECL (emitter-coupled logic) also known as *CML* (current-mode logic). This family is based on a differential amplifier which is basically an anolog device. Nevertheless it has important application in digital logic and is the faster operating of all the logic families with delay times as low as 1 nanosecond per gate.

Flip-Flops

A flip-flop is a bistable circuit and another important element in digital logic. Since it is capable of storing *one* bit of information it is functionally a *1-bit memory unit*. Because this information is locked or 'latched' in place, a flip-flop is also known as a *latch*. A combination of n flip-flops can thus store an *n-bit word*, such a unit being referred to as a *register*.

A basic flip-flop circuit is formed by cross-coupling two single-input NOT gates, the output of each gate being

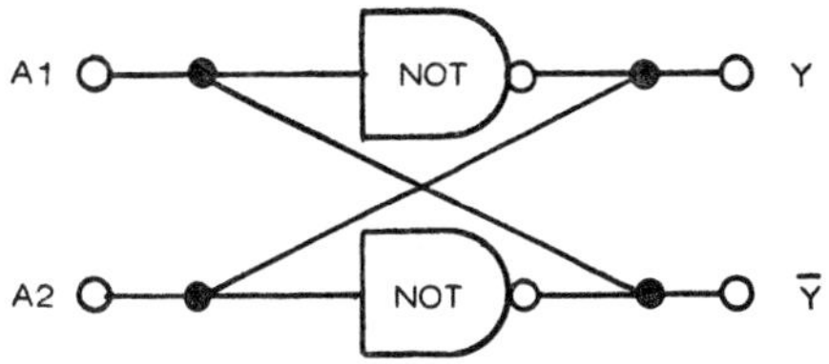

11.6 1-bit memory or latch circuit obtained by cross-coupling two NOT gates (or two single-input NAND gates). The output has two states $Y=1$, $\overline{Y}=0$; or $Y=0$, $\overline{Y}=1$. For flip-flops the symbols Q and $\overline{Q}$ are often used for the outputs instead of Y and $\overline{Y}$ respectively.

connected back to the input of the other gate—Fig. 11.6. However, to be able to preset or clear the state of the flip-flop, two two-input NOT gates cross-coupled are necessary, each preceded by single-input NOT gates as shown in Fig. 11.7.

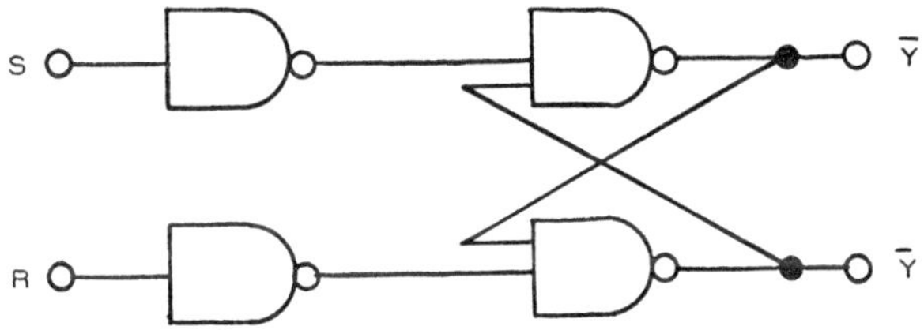

11.7 Flip-flop circuit with preset using four NOT gates. S is the set or preset input. R is the reset or clear input.

When the flip-flop is used in a pulsed or clocked system the preceding gates are known as the *steering* gates with the cross-coupled two-input gates forming the *latch*. This particular configuration is also known as a S-R or R-S flip-flop.

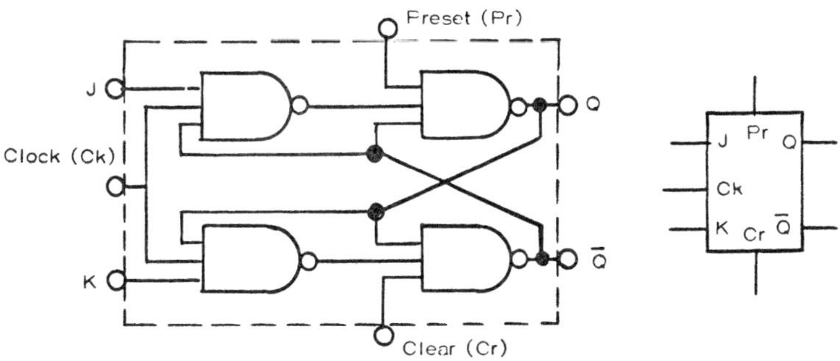

11.8 J-K flip-flop circuit (*left*) with corresponding symbol (*right*).

Two other variations of the flip-flop are also produced as integrated circuits:
J-K flip-flop—which is an S-R flip-flop preceded by two AND gates. This configuration removes any ambiguity in the truth table. It can be used as a T-type flip-flop by connecting the J and K inputs together (*see* Fig. 11.8 for connections).
D-type flip-flop—which is a J-K flip-flop modified by the addition of an inverter (*see* Fig. 11.9). It functions as a 1-bit delay device.

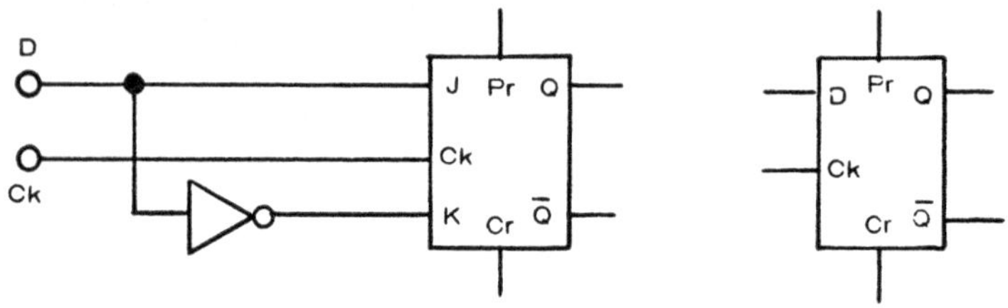

11.9 A D-type flip-flop circuit (*left*) is provided by a J-K flip-flop allied to an inverter. The symbol for a D-type flip-flop is shown on the right.

Fan-in and Fan-out

The terms *fan-in* and *fan-out* are used with IC logic devices. Fan-in refers to the number of separate inputs to a logic gate. Fan-out is the number of circuit loads the output can accommodate, or in other words the number of separate outputs provided. Fan-out is commonly 10, meaning that the output of the gate can be connected to 10 standard inputs on matching gates. Each separate input represents a load, the higher the number of separate loads the higher the current output of the device providing fan-out needs to be in order to provide the *standard load* on each input, i.e. passing enough current to drop each input voltage to the design figure.

It is possible to increase fan-out by replacing diode(s) with transistor(s) in the device concerned, so 10 is by no means a maximum number.

ROM

ROM stands for read-only memory, a system capable of converting one code into another. The best known application is to convert the reading of a digital instrument such as an electronic calculator into a numerical read-out via an LED (light emitting diode) display. The advantage of a ROM is that it is programmable and thus adaptable to different read-out systems. It does not follow, however, that it uses the minimum number of components to match a particular application. Special IC chips designed for a specific application may be more economic in this respect, but not necessarily in cost, unless there is a very large demand for that particular IC. The calculatờr market is a case in point where a special chip can offer advantages over a ROM.

RAM

RAM stands for random-access memory and is basically a collection of flip-flops or similar devices capable of memorizing information in binary form. Information can be written-in or read out in a random manner.

The Shape of Digital ICs

In physical appearance, most digital ICs look like any other dual in-line (or sometimes quad in-line) IC package, or ceramic flat packages. They are not readily indentified as digital ICs, therefore, (except by type number) although their function is quite specific. The more complicated digital ICs may, however, have considerably more pins than usual. It is also common practice to give pin diagrams which not only definethepinpositionsbutalsotheirspecificfunction(Fig.11.10).

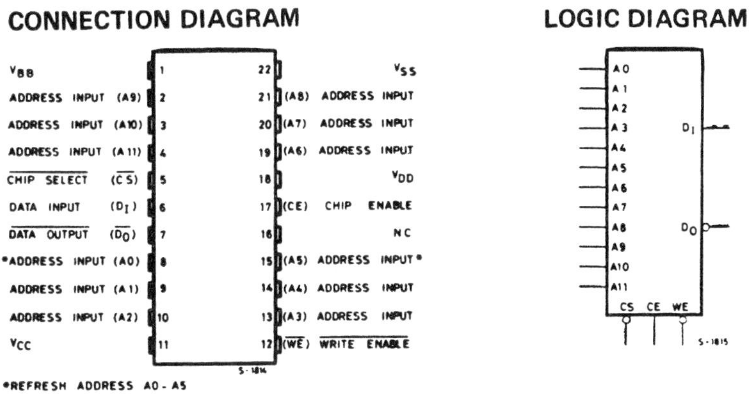

11.10 Example of a Random Access Memory integrated circuit with connection diagram (Mullard M340 with a capacity of 4096 'bits'). This is in the physical form of a 22 lead dual-in-line package.

Chapter Twelve

ELECTRONIC ORGANS

ONE OF THE MAJOR PROBLEMS in the design of electronic organs is the large number of mechanical contacts called for using conventional (discrete component) circuitry. With two manuals of four octaves each, for example, 98 mechanical contacts are needed. This not only complicates construction but could also be a source of trouble in operation. There is often the limitation that each key is only able to play one note. It is desirable for electronic organs to be able to play more than one octave-related note per key, increasing the number of mechanical contacts required by that factor, e.g. $5 \times 98 = 490$ contacts for the example quoted to be able to play five octave-related notes per key.

A number of integrated circuits have been developed, usually based on digital logic, to overcome such limitations. Many also provide additional features which may be desirable. An example is the (Mullard) TDA1008 which consists of a matrix of gate circuits with eight divide-by-two gates in each circuit. It is a 16-lead dual-in-line plastic package (SOT-38).

One drive input only is required for delivering nine octave-related notes and, by actuating a key input, five successive signals out of the nine can be selected and transferred to the output. Five key inputs are available, each selecting a different combination. Other features which are available are 'sustain' and 'percussion' of the output signals; and also 'decay' of modulations.

Further simplification of an electronic organ circuit can also be provided by using a top octave synthesizer (TOS) instead of a series of master oscillators to derive the twelve top octave frequencies required for a 'full' organ. A TOS must be associated with a *master oscillator* capable of generating a suitable 'least common multiple' frequency, with the TOS following it, then providing the twelve highest notes. Used with a suitable gating matrix, further sub-multiples of these notes are obtained, e.g. in the case of the TDA1008 the following output frequencies are available from the five keys, where f is the actual input frequency:

	key 1	key 2	key 3	key 4	key 5
output 1	f	f/2	f/4	f/8	f/16
output 2	f/2	f/4	f/8	f/16	f/32
output 3	f/4	f/8	f/16	f/32	f/64
output 4	f/8	f/16	f/32	f/64	f/128
output 5	f/16	f/32	f/64	f/128	f/256

This, in effect gives nine different notes available from each of twelve available input frequencies from the TOS, or 96 different notes. Further, operating two or more keys simultaneously will give the sum signal of these frequencies.

Master Oscillator

A suitable frequency for the master oscillator is about 4.5 MHz. A variety of circuits can be used providing they have suitable stability and the necessary amplitude and slew rate for driving the TOS properly. If the master oscillator is a sine wave generator, then it will be necessary to follow this with a Schmitt trigger to obtain the required slew rate. This is not necessary with a square-wave generator and a very simple circuit of this latter type based on the NAND gates contained in the HEF4011 integrated circuit is shown in Fig. 12.1. This requires a

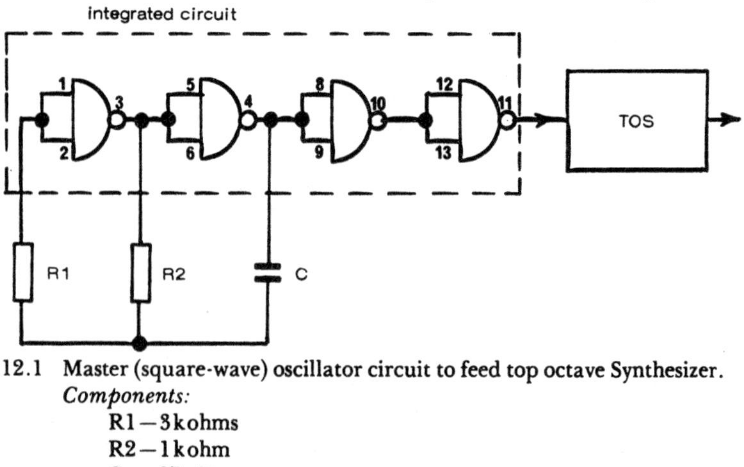

12.1 Master (square-wave) oscillator circuit to feed top octave Synthesizer.
Components:
R1—3 kohms
R2—1 kohm
C —27 pF
IC—HER4011
TUS—AYE0214

stabilized 12-volt supply, as does the TDA1008, so the same supply can be used for both the master oscillator and TDA1008.

The master oscillator output connects to the Top Octave Synthesizer, the tone outputs of which form the input to the TDA1008. They can be directly connected since the input signal pin of the TDA1008 has an impedance of at least 28 kohms.

Gate Matrix

Connections to the TDA1008 integrated circuit are shown in Fig. 12.2. The different levels of supply voltage required are 6

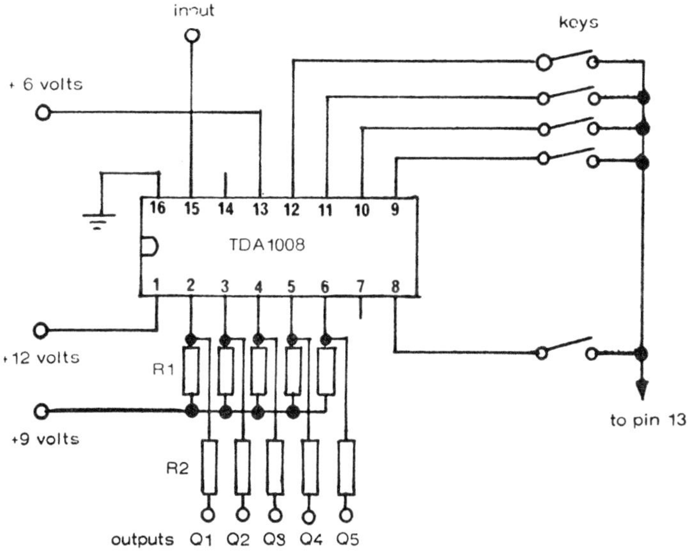

12.2 Basic electronic organ circuit using five keys. Resistors R1 are all 1 kohm. Resistors R2 are all 100 kohms. Q1, Q2, Q3, Q4 and Q5 are the tone outputs to feed an audio amplifier circuit with loudspeaker.

volts, 9 volts and 12 volts, as shown. The five keys can be directly connected, although current-limiting resistors can be used in each key line if necessary.

Five different output frequencies are available at each output Q1, Q2, Q3, Q4, Q5, depending on which key is activated (*see* table above). To avoid sub-harmonics being generated it is advisable to connect any not-required Q outputs to the + 6 volt supply line.

Sustain

To actuate sustain and percussion effects, a time-delay circuit can be added associated with each key, as shown in Fig. 12.2. This circuit will sustain the tone(s) for a period after

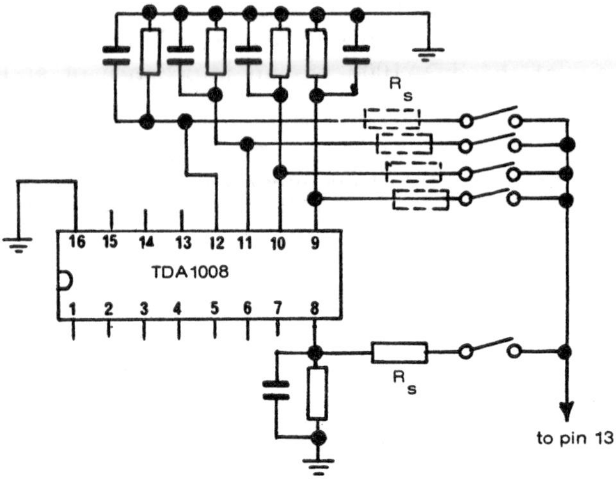

12.3 'Sustain' added to the circuit of Fig. 12.2. Other components are connected as before
Component values:
Resistors—2.2 M ohms
Capacitors—0.5 μF
R_s —Series resistors, if required

release of the key, but with the resistor also providing a certain delay time. The addition of a series resistor (RS) will delay the build-up of notes, depending on the RC time constant of this resistor and the associated capacitor in the circuit. Component values given are selected for good tonal response, but this is also a matter of personal preference and so some adjustment of values may be preferred. It is also possible to shorten the decay time of the sustain by adjusting the voltage applied to pin 7. A circuit for doing this is shown in Fig. 12.4.

Percussion

If percussion is required this can be arranged by connecting a capacitor to pin 8 to discharge during keying, associated with a series resistor to give a suitable time constant. Using a 0.47 F capacitor, a suitable series resistor value can be found by experiment. The decay time is also adjustable via the circuit shown in Fig. 12.4.

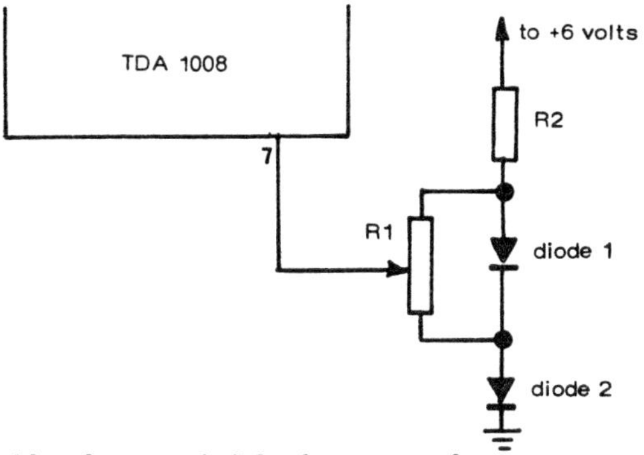

12.4 Adjustable voltage to pin 7 for decay control.
Component values:
R1 — 100 ohms potentiometer
R2 — 100 ohms
diode 1 — BZX75 C2V1
diode 2 — BOW62

To retain sustain as well, the circuit shown in Fig. 12.5 should be used. If sustain is wanted, switch S1 is closed and

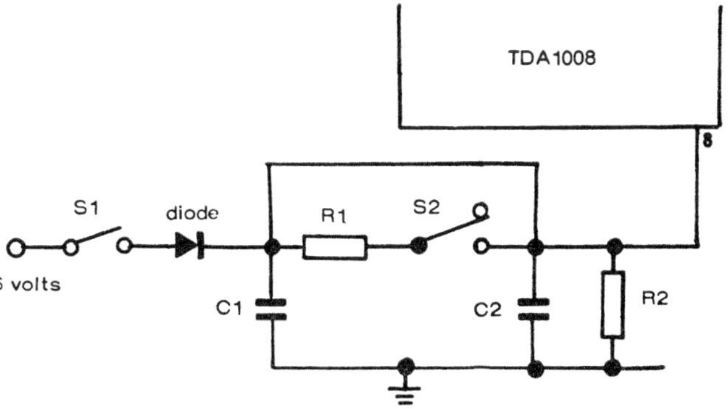

12.5 Percussion circuit with sustain, connecting to pin 8.
Component values:
Resistors
R1 — 10 kohms
R2 — 2 M ohms
Capacitors
C1 — 0.47 μF
C2 — 0.47 μF

switch S2 opened. C1 then remains charged to sustain the note as long as a key is held down. Once the key is released the note will decay at the rate established by the decay circuit connected to pin 7. To operate percussion, switch S1 is open and switch S2 closed.

Chapter Thirteen

MISCELLANEOUS CIRCUITS

HI-FI TONE CONTROLS

Tone controls fitted to domestic radios and equivalent circuits are seldom of high quality. This does not usually matter for AM reception (which can never be Hi-Fi); but can degrade the performance on FM reception. Similar remarks apply to the tone controls fitted to lower priced record players and tape recorders.

High quality tone controls generally demand quite complex circuits. ICs enable the number of discrete components required to be substantially reduced and, at the same time, offer other advantages such as a high input impedance which matches a typical high impedance source. Tone control can also be combined with audio amplification in IC circuits.

Fig. 13.1 shows a complete circuit based around a TCA8305 integrated circuit incorporating a feedback network which attenuates the low frequencies and boosts the high frequencies. At the same time high frequencies can be attenuated by the treble control potentiometer at the input. The volume control, also on the input side, provides 'loudness control' at both high and low frequencies to compensate for the loss of sensitivity of the human ear to such frequencies (i.e. both high and low frequencies tend to sound 'less loud' to the ear).

A simpler circuit, using the same IC, is shown in Fig. 13.2. This has a single tone control potentiometer. The circuit provides flat response at middle frequencies (i.e. around 1 kHz), with marked boost and cut of up to ± 10 decibels at 110 Hz and 10 kHz respectively in the extreme position of the potentiometer.

A (Baxandall) Hi-Fi tone control circuit associated with another type of op-amp is shown in Fig. 13.3. The IC in this case is the CA3140 BiMOS op-amp. The tone control circuit is conventional and only few additional discrete components are required to complete the amplifier circuit around the IC. This circuit is capable of ± 15 decibels bass and treble boost and cut

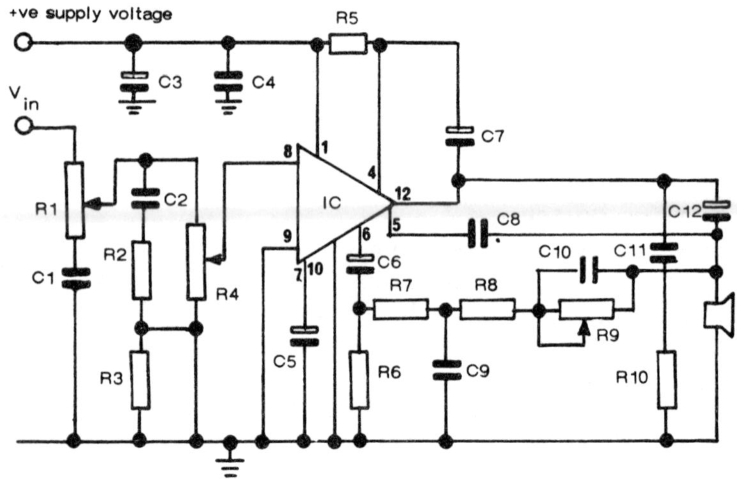

13.1 Hi-Fi tone control circuit suitable for receivers, record players and tape recorders and charaterized by a high input impedance. Potentiometer R1 is the treble control. Potentiometer R9 is the bass control. Potentiometer R4 is the volume control.

Component values:

R1—47kohms log pot	C1— 47nF
R2—10kohms	C2— 820pF
R3—1.8kohms	C3— 100μF
R4—100kohms log pot	C4— 0.1μF
R5—100ohms	C5— 100μF
R6—15ohms	C6— 250μF
R7—470ohms	C7— 100μF
R8—470ohms	C8— 100pF
R9—25kohms log pot	C9— 0.33μF
R10—1ohm	C10— 0.22μF
	C11— 0.1 μF
IC—TCA8305	C12— 1000μF

at 100Hz and 10kHz respectively.

An alternative circuit using the same IC and giving a similar performance is shown in Fig. 13.4. Both of these circuits require a supply voltage of 30-32 volts. Fig. 13.5 shows the same two circuits modified for dual supplies.

LED DISPLAY BRIGHTNESS CONTROL (Fig. 13.6, page 113)

How well an LED shows up is dependent on the ambient light falling on it. In dim light the display is usually quite bright. In

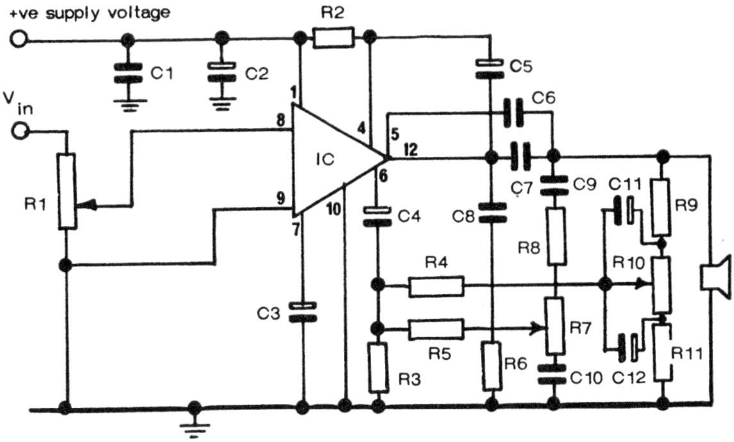

13.2 Alternative Hi-Fi tone control circuit with separate high and low frequency feedback. Potentiometer R1 is the volume control. Potentiometer R7 is the treble control and potentiometer R10 the bass control.

Component values:

R1—100 kohms log pot	C1— 0.1μF
R2—100 ohms	C2— 100μF
R3—18 ohms	C3— 100μF
R4—180 ohms	C4— 500μF
R5—27 ohms	C5— 100μF
R6—1 ohm	C6— 82pF
R7—10 kohms log pot	C7— 1000μF
R8—150 ohms	C8— 0.1μF
R9—330 ohms	C9— 0.15μF
R10—10 kohms log pot	C10— 2μF
R11—15 ohms	C11— 1μF
IC—TCA8305	C12— 2.2μF

direct sunlight it may be difficult to see at all. The circuit shown in Fig. 13.6 provides an automatic brightness control of a (single) LED by using a silicon photodiode to sense the amount of ambient light and feed a proportional signal to the TCA315 op-amp integrated circuit. As the intensity of light increases the output current from the op-amp increases in proportion, and vice versa, thus automatically compensating the brightness of the LED for artifical light in an inverse manner. The brighter the ambient light the brighter the LED glows, and vice versa.

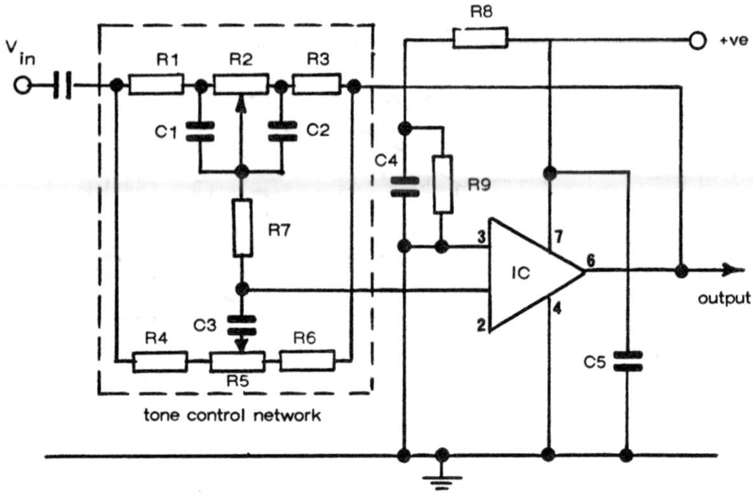

13.3 Simple Hi-Fi tone control circuit. Component values are determined for a supply voltage of 32 volts. Potentiometer R2 is the bass control. Potentiometer R5 is the treble control. Components within the dashed outline comprise the tone control network.

Component values:

R1—240 kohms	C1—750 pF
R2—5 M ohm log pot	C2—750 pF
R3—240 kohms	C3— 20 pF
R4—51 kohms	C4—0.1 μF
R5—5 M ohm linear pot	C5—0.1 μF
R6—51 kohms	Coupling Capacitor
R7—2.2 M ohm	(C8)—0.047 μF
R8—2.2 M ohms	
R9—2.2 M ohms	
IC—CA3140	

The potentiometer (R6) is used for setting up the circuit initially. With a 2.5 volt supply, and with the photodiode in complete darkness, R6 should be adjusted to give a current reading of about 100 μA (0.1 milliamps), using a meter in one battery lead to check. With this adjustment, and the type of photodiode specified, the LED will then receive an impressed current of 5 mA per 1000 lux illumination of the photodiode.

Components:

Integrated circuit TCA315 op-amp
Photodiode BPW32

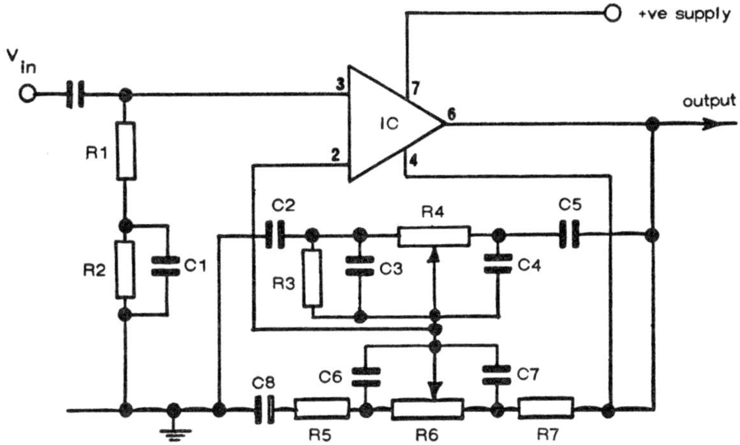

13.4 Another Hi-Fi tone control circuit. Potentiometer R4 is the treble control. Potentiometer R6 is the bass control. Supply voltage is 30 volts.

Component values:

R1 – 5.1 Mohms	C1 – 0.1 μF
R2 – 2.2 Mohms	C2 – 0.01 μF
R3 – 18 kohms	C3 – 100 pF
R4 – 200 kohms linear pot	C4 – 100 pF
R5 – 10 kohms	C5 – 0.001 μF
R6 – 1 Mohm log pot	C6 – 2 μF
R7 – 100 kohms	C7 – 0.002 μF
IC – CA3140	C8 – 0.005 μF
IC – CA3140	

LED LD30 (or equivalent)

Resistors: R1 47 k ohm
R2 47 k ohm
R3 220 ohm
R4 47 ohm
R5 10 M ohm
R6 250 k ohm potentiometer

LED RADIO TUNING SCALE (Fig. 13.7, page 114)

This simple circuit displays the tuned frequency of a radio in terms of spots of light instead of (or in addition to) the usual pointer moving over a scale. An array of 16 LEDs should be sufficient to indicate station positions with suitable accuracy

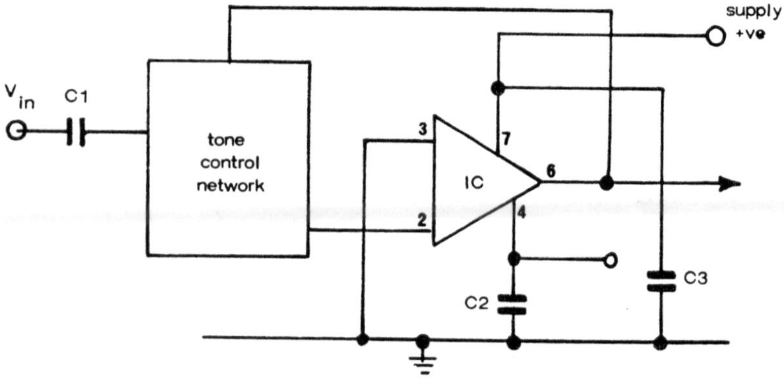

13.5 Tone control for dual supplies. The tone control network is the same as that in the dotted outline of Fig. 13.3. Supply voltage is 15 volts.
Component values:
C1 – 0.047 μF
C2 – 0.1 μF
C3 – 0.1 μF
IC – CA3140

over a typical medium frequency waveband (i.e. 520 kHz to 1600 kHz). The display is driven by a Siemens UAA170 integrated circuit. A phototransistor is also used to match the brightness of the display automatically to ambient light intensity, i.e. dimming the display in dull light and brightening the display to make it clearly visible in sunlight.

The complete circuit is shown in Fig. 13.7. The UAA170 is controlled via the voltage divider formed by R1 and R2 supplying the tuning voltage for the AM tuning diode incorporated in the IC. Since this diode has non-linear characteristics, stations on the left (lower frequency) end of the tuning scale will be more closely concentrated, consistent with station spacing on this broadcast band.

The circuit will work on most normal transistor radio supply voltages (i.e. V_S = 10 to 18 volts), and with an input voltage for frequency indication of V_S = 1.2 to 27 volts using two (Siemens) LD468 LED-arrays. Voltage at the divider point between R1 and R2 should be between 0.06 and 1.16 volts and can be adjusted by R1 if necessary. The actual brightness of the

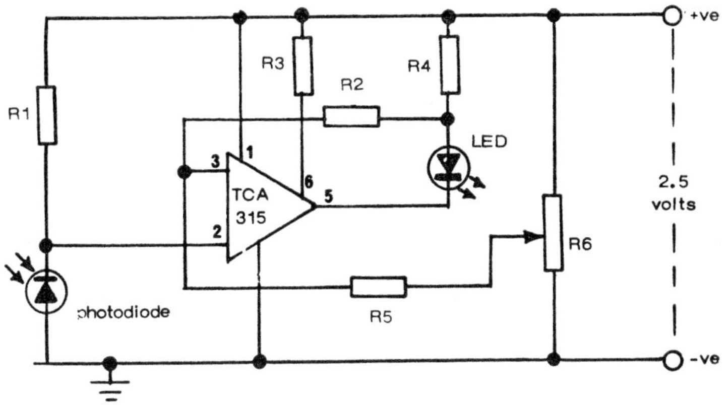

13.6 Circuit for automatic control of brightness of an LED using a photodiode to sense the level of illumination.

display is automatically controlled by the phototransistor BP101/1, and is also adjustable via the 1 kohm potentiometer.

CAR THIEF ALARM (Fig. 13.8, page 115)

This is another circuit originated by Siemens and based around their TDB0556A dual timer IC. The first timing circuit of this device is used as a bistable multivibrator with the circuit activated by switch S1. Output level remains at zero, set by the voltage applied to the threshold input pin 2 until one of the alarm contact switches is closed causing C1 to discharge.

'Press-for-off' alarm switches can be fitted to the doors, bonnet and boot lid, so arranged that opening of a door or lid completes that switch contact. This will produce an output signal held for about 8 seconds, pulling in the relay after an initial delay of about 4 seconds. The horn circuit is completed by the relay contacts so the horn will sound for 8 seconds. After this the relay will drop out (shutting off the horn) until capacitor C1 charges up again. This will take about 3 seconds, when the relay will pull in once more and the horn will sound again. This varying signal of 8 seconds horn on, 3 seconds horn off, will be repeated until switch S1 is turned off (or the battery is flattened). This type of alarm signal commands more attention than a continuous sounding alarm such as can be given by straightforward on-off electrical switching.

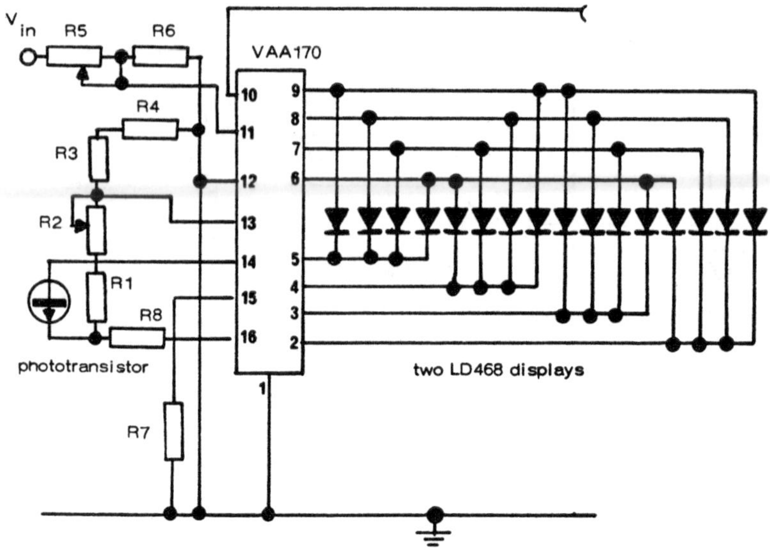

13.7 Sixteen LED display to replace or augment the usual pointer and scale indication of tuned frequency on an AM radio receiver.
Component values:
R1 – 330 kohms
R2 – 1 kohm potentiometer
R3 – 6.8 kohms
R4 – 2.7 kohms
R5 – 10 kohms potentiometer
R6 – 470 ohms
ICA – UAA170
phototransistor – BP101/1
LED – two LD468 displays

The complete circuit is shown in Fig. 13.8 with suitable component values, wired in to appropriate points on a car electrical system.

INTERCOM

The TCA830S is a powerful, inexpensive op-amp IC which makes it a particularly attractive choice for intercoms since the circuit can be built with a minimum number of components. Many other op-amps do not produce the power required for loudspeaker operation without the addition of a further stage of transistor amplification. The basic circuit is contained at the 'main' station when the 'distant' station merely comprises a

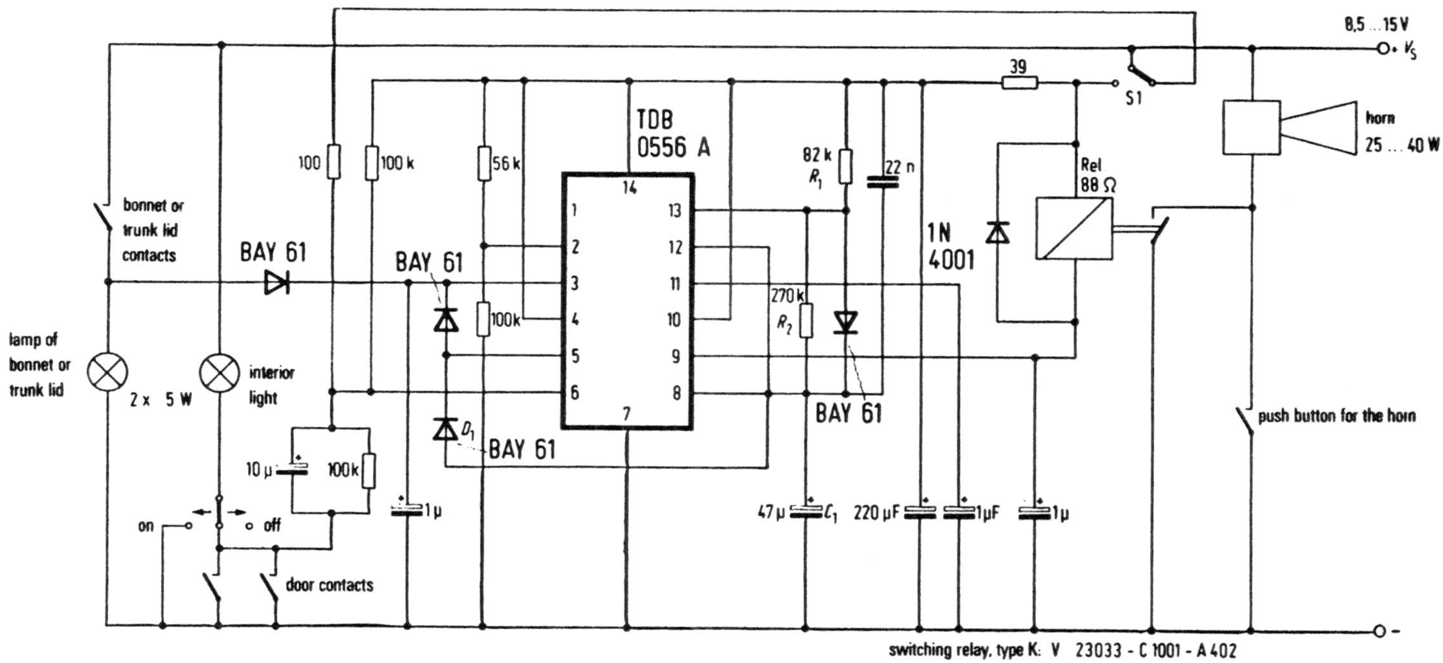

13.8 Circuit design by Siemens for a car thief alarm. The IC is a TDB-0556A. All component values are shown on the diagram.

loudspeaker and a 'calling' switch. The two stations are connected by a 3-wire flex.

The circuit is shown Fig. 13.9. The TCA830S requires a heat sink and is fitted with tabs. A printed circuit is recommended,

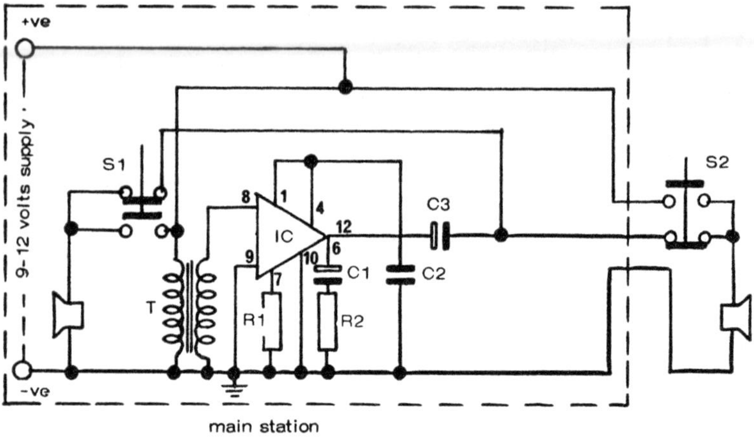

13.9 Intercom circuit using the TCA830S integrated circuit. This IC is powerful enough to operate fairly large loudspeakers. Component values are given in the text.

incorporating two 1 in. (25 mm) squares of copper to which the IC tabs can be soldered for the heat sink. Component positioning is not critical since the circuit handles only audio frequencies.

The transformer (T) has a 50:1 turns ratio and is used as a step-down transformer between the IC and speaker(s)—also working as a step-up transformer between speaker(s) and IC for working in the reverse mode. In other words the transformer coil with the larger number of turns is connected to pin 8 on the IC. Instead of purchasing this transformer ready-made it can be wound on a stack of standard transformer core laminates 0.35 mm thick, giving a core cross-section of 22.5 mm^2. Windings are 600 turns of 0.2 mm (36 s.w.g.) and 300 turns of 0.06 mm (46 s.w.g.) enamelled copper wire.

The purpose of the transformer is to enable standard 4 to 16-ohm loudspeakers to be used both as microphones and speakers. These speakers can be of any size, bearing in mind that the maximum power output of the circuit is of the order of 2 watts on a 12-volt supply. The intercom circuit will work on any battery voltage down to 6 volts, 9 or 12 volts being recommended for general operation.

Components:
(SGS—ATES) TCA8305 integrated circuit
Resistors R1 20kohms
R2 29ohms
Capacitors C1-100μF electrolytic 3V
C2 0.1μF
C3 1000μF electrolytic 12V
Transformer (T) 50:1 turns ratio, power rating 5W.
Loudspeaker 4ohms (preferred)
Switch S1: press break/make
S2: press make/break

ICE WARNING INDICATOR

This very simple circuit uses a thermistor as a temperature sensor together with three CA3401E op-amps and a minimum of external components. The operating point of the circuit is set by the potentiometer (R2) so that, at an ambient air temperature approaching freezing point, the light emitting diode (LED) starts to flash. As the temperature falls the rate of flashing increases until the LED glows continuously once freezing point is reached. Accurate calibration can be carried out in the freezer compartment of a domestic refrigerator with the door open, in conjunction with a thermometer.

The complete circuit is shown in Fig. 13.10. IC1, IC2 and IC3 are separate op-amp circuits contained in the IC. Thus pins 1 and 6 are the input to IC1 and pin 5 the output of IC1; pins 11 and 12 the input to IC2 and pin 10 the output of IC2; and pins 2 and 3 the input to IC3 and pin 4 the output of IC3. Pins 8, 9 and 13 are ignored. Pin 7 connects to the earth side of the circuit; and pin 14 to battery plus side.

Layout of this circuit is not critical but all component leads should be kept as short as possible and the LED located some distance away from the integrated circuit. This circuit is powered by a 12 volt battery.

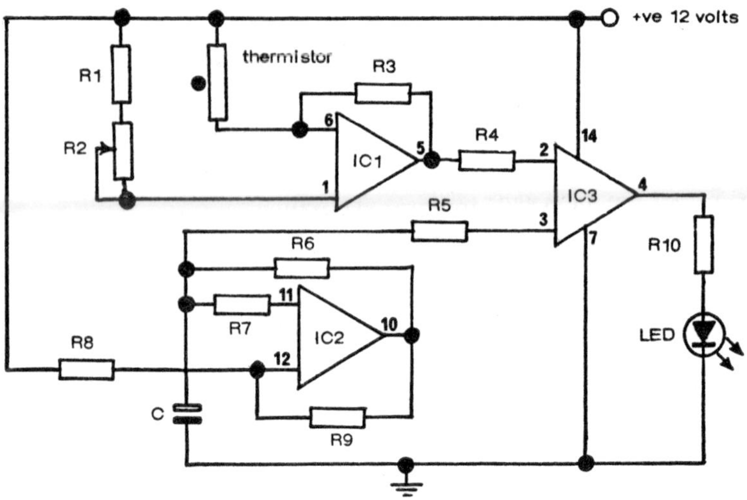

13.10 Circuit for an ice-warning indicator. Adjustment of potentiometer R2 can set the circuit to flash the LED as air temperature approaches freezing point, with LED staying permanently alight once freezing temperature is reached.

Component values:

R1 – 33 kohms	IC1, IC2, IC3 – CA3401E
R2 – 20 kohm potentiometer	LED – light emitting diode
R3 – 150 kohms	Thermistor – Mullard VA1066S (or equivalent)
R4 – 3 Mohms	
R5 – 3 Mohms	
R6 – 30 kohms	
R7 – 3 Mohms	
R8 – 10 Mohms	
R9 – 10 Mohms	

DIGITAL VOLTMETER

A digital voltmeter (known as a DVM) has several advantages over a conventional pointer-and-needle meter, for example:

1. Easier reading with direct presentation of reading in digits.
2. Greater accuracy and high speed of reading.
3. Higher sensitivity.
4. Greater resolution.

Unfortunately the circuitry required for a DVM is quite complicated, making it much more expensive than its simple analog counterpart in the form of moving coil instrument. However, by using ICs the necessary circuitry for a DVM can be simplified and miniaturized and is within the scope of the amateur to build. The following design by Siemens avoids the use of expensive components and its performance is comparable with that of ready-made DVMs in the medium-price range (well over £100!). It has a basic range of up to 9.9 volts with an accuracy of better than 99 per cent.

The complete circuit is shown in Fig. 13.11. The input voltage is converted to a proportional frequency by the op-amp TBA221 connected as an integrating amplifier and the following monostable multivibrator TDB556A (IC2). The resulting output pulse (at pin 5 of IC2) is determined by the time constant of R4 and C4 and is of the order of 1.5μs. This pulse turns transistor T1 'on' and 'off', the multivibrator thus supplying pulses to the clock input of the counter SAJ341 with a repetition frequency proportional to the input voltage.

These pulses are counted during a measuring interval defined by the other half of the astable multivibrator TDB556A (IC1) with a duty cycle of <0.5. Its output directly controls the blocking input of the counter (SAJ341). At the beginning of each measuring interval, 5AJ341 is reset to Q_A, Q_B, Q_C, Q_D = L (corresponding to decimal 0) by a short L-pulse applied to the reset input IR. This reset pulse is produced by the measuring-interval generator, the inverting transistor T2 and the following differentiation circuit.

The display, which can be extended to four digits, operates on a time-multiplex basis using a level converter (TCA671), decoder (FLL121V) and display driving transistors BC307 and BC327.

The circuit is set up using a known input voltage (preferably between 2 and 3 volts). Potentiometer R1 is then adjusted to show the correct reading on the display. If this is not possible then the value of resistor R2 should be changed for the next nearest value up or down, i.e. 270 or 180 kilohms as found appropriate (one value will make matters worse, the other better).

The circuit needs two separate power supplies of +5 volts at 300 milliamps and -12 volts at 200 milliamps. For accurate working of the meter both supply voltages should be regulated.

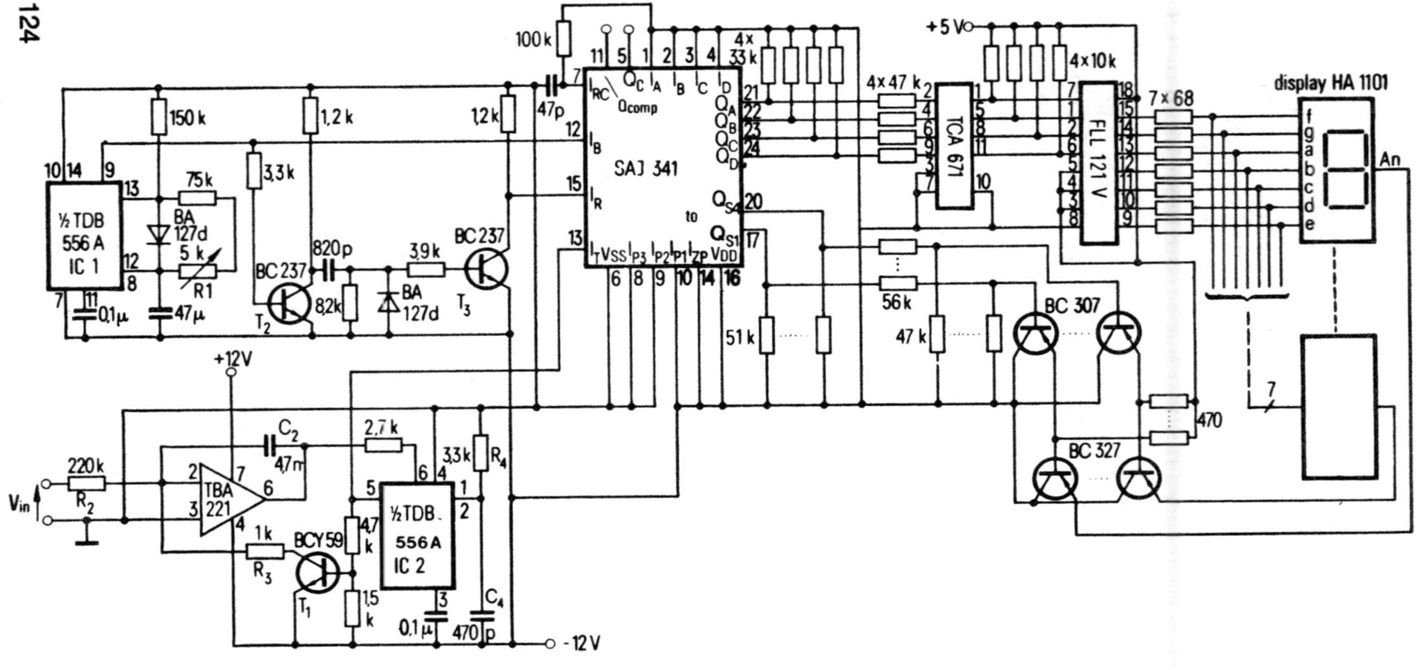

13.11 Digital voltmeter circuit (Siemens). All component values are marked on the drawing.

Components:
IC1 & IC2 – Siemens TDB0556A
Siemens TBA221 integrating op-amp
Siemens SAJ341 universal converter
Siemens TCA671 level converter
Siemens FLL121V decoder
Note: all the above are integrated circuits.
HA1101 seven-sequence LED display
Capacitor and resistor values are shown on the circuit diagram.
Diode and transistor type numbers are shown on the circuit diagram .

INFRA-RED TRANSMITTER AND RECEIVER

There are three practical possibilities for remote control signalling: radio (as in model radio control systems); ultrasonics; and light transmission. The latter is the simplest in terms of components and circuitry, especially where simple on/off command only is required. It can be extended to more channels, but at the expense of more complicated circuitry.

Using infra-red light transmission it is possible to achieve a range of 100 feet (30 metres) or more quite readily in normal ambient light. Even greater range is possible if the transmitter light beam is focused by a simple lens system. Such infra-red remote control systems have become highly practical with the appearance of high-efficiency LEDs with a high infra-red transmission and suitable photodiodes which can be used as detectors in receivers. As with other remote control systems the basic units involved are a *transmitter* and *receiver*.

Single-channel infra-red transmitter

This circuit uses the Siemens GoAs improved light-emitting diode LD241 in a pulse modulated transmitter circuit involving the use of two oscillators, a sub-carrier frequency of 50 kHz modulated by a frequency of 10 Hz, the second oscillator having a duty cycle of 250:1. These circuits are based around four CMOS NAND-gates (available in a single IC). The LED is square-wave modulated by a Darlington pair of NPN transistors.

The complete transmitter circuit is shown in Fig. 13.12 and is quite straightforward. Despite drawing a peak current of 1

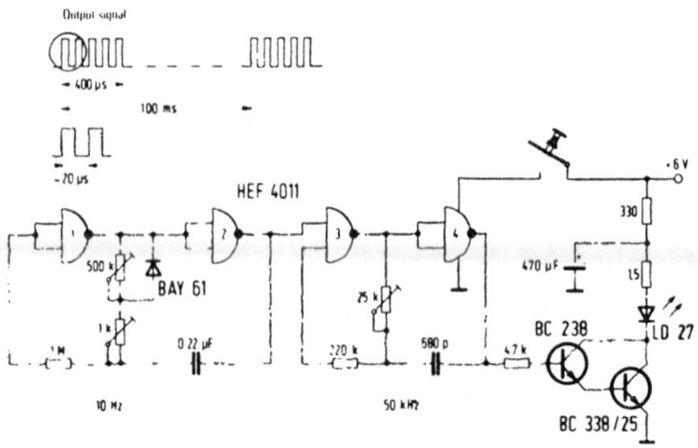

13.12 Design for an infra-red transmitter (Siemens). Component values are shown on the diagram, but a complete specification for the active components is:
IC1, IC2, IC3, IC4 – 4 × CMON NAND-gates type (Siemens) HEF4011P
Transistors – BC238/25 (or equivalent)
LED – Siemens LD27
diode – BAY61

amp the average current drain is only 2 mA with a 6-volt battery supply, the peak current actually being supplied from the 470 μF capacitor. This is possible since the 5 kHz output pulse train has a duration of only 400 μs in a repetition period of 100 ms.

Single-channel Infra-red Receiver.

By comparison the receiver circuit is more complex since it employs six discrete transistor plus a Darlington pair in addition to three NAND-gates – Fig. 13.13. The detector is a BPW34 photodiode matched to an input impedance of 80 k ohms at 50 kHz. Signals are received in the form of an infra-red pulse train from the transmitter. The receiver circuit following the photodiode amplifies, clips and rectifies the pulse train signal and applies it to a monostable multivibrator which covers the space between two pulse trains. This means that a *dc* voltage is available at the output of the receiver as long as the transmitter signal is held on. This receiver output can be used to operate a relay, simple escapement or a signalling light (e.g. a filament bulb or LED).

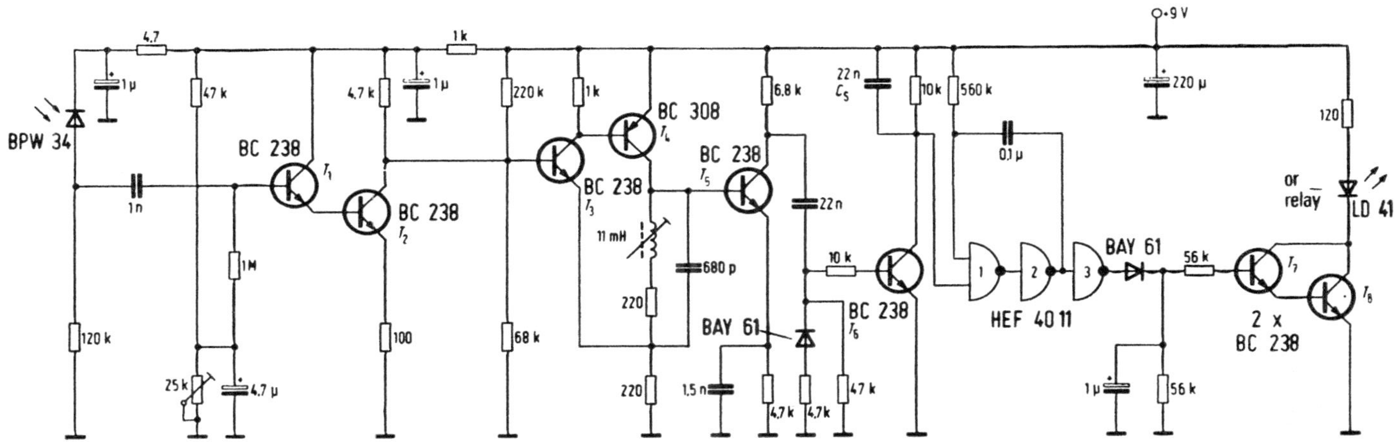

13.13 Design for a matching infra-red receiver (Siemens). Component values are given on the diagrams. A complete specification for the active components is:
IC—HEF4011P
photodiode—BPW39
transistors—7 × BC238 (or equivalent)
1 × BC308 (or equivalent)
LED—LD41/A
diodes—BAY61
inductance—11 mH

Since ambient light will introduce a 'noise' voltage in the diode or interference, the circuit is intended for narrow band working which operates by placing an infra-red filter in front of the photodiode. This can be an infra-red photographic filter, or a section of unexposed but developed colour film (e.g. Agfa CT18). The transmitter-receiver combination should then work satisfactorily in ambient light intensities up to 10000 lux with fluorescent light, 4000 in sunlight, or 500 lux maximum in the case of filament lighting.

A simpler receiver circuit is shown in Fig. 13.14 but will only be suitable for working in dull ambient light (less than 500 lux).

ELECTRONIC REV COUNTER

The (Mullard) SAK140 is an integrated circuit designed as a revolution counter for car engines, etc. Connected to the contact breaker it is fed by input pulses at 'engine speed' rate and converts these pulses into output current pulses of constant duration and amplitude. The output pulse duration is determined by an external Resistor—Capacitor network. By suitable choice of R and C, the pulse 'count' can be indicated on any milliameter. The circuit will also work on any supply voltage between 10 and 18 volts (e.g. from a car's 12-volt battery) and performance is independent of actual supply voltage (or variation in supply voltage).

The complete circuit is shown in Fig. 13.15. Resistor R1 is selected so that the input current does not exceed 10mA (a suitable value for 12-volts supply is 15kohms, when typical input current will be 5mA). The diode acts as a voltage regulator to prevent overloading by large input pulses.

The peak output current is determined by the value of R2 plus R3. This should be at least 50ohms, the actual value being chosen to suit the range of the milliameter used. If R2 is made 50 ohms, then R3 can be made 1 kohm, say, and adjusted to suit the range of the milliameter.

The output pulse duration is determined by the combination of R4 and C2. Suitable values can be found by experiment, the suggested starting point being:

R4—270kohms
C2—10nF

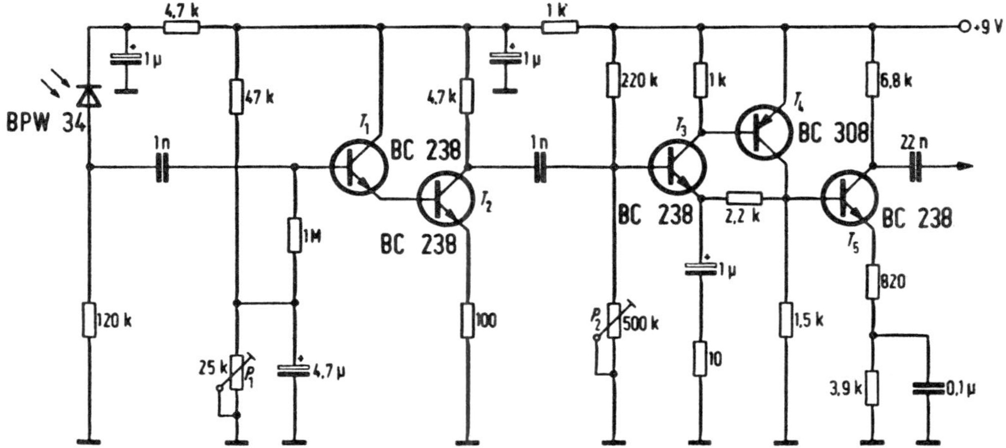

13.14 Simpler broadband infra-red receiver circuit (Siemens). Component values are given on the diagrams. A specification for the active components is:
Transistors—4 × BC238 (or equivalent)
photodiode—BPW34

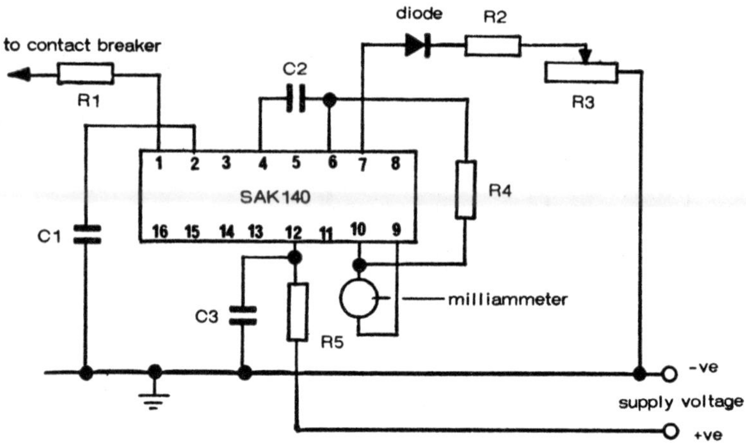

13.15 Electronic rev counter circuit using the SAK140 IC.

Component values:

Resistors	Capacitors
R1 – 15 kohms	C1 – 220 nF
R2 – 50 ohms	C2 – *
R3 – 1 kohms	C3 – 100 nF
R4 – *	
R5 – 27 ohms	
IC – SAK140	**see* text

QUARTZ CRYSTAL CLOCK

The (Mullard) SAA1114 is a C-MOS integrated circuit designed to work as the 'heart' of a crystal controlled clock powered by a single battery. It comprises a master 4 MHz oscillator, a 22-stage frequency divider and a driver for a unipolar stepper motor. With a crystal frequency of 4,1943 MHz, the output is in the form of a 1 Hz (1 second) pulse of 31.25 milliseconds duration.

A complete clock circuit is shown in 13.16 and requires only a few external components. The quartz crystal is a critical component and is associated with a trimmer capacitor C1 for time adjustment. Maximum supply voltage is 3 volts, the circuit drawing a current of about 50A and supplying a motor output current of about 50 mA.

Another version of this particular IC is also available which incorporates an alarm circuit triggered by an alarm switch operated by the clock hand movement. Output of this alarm

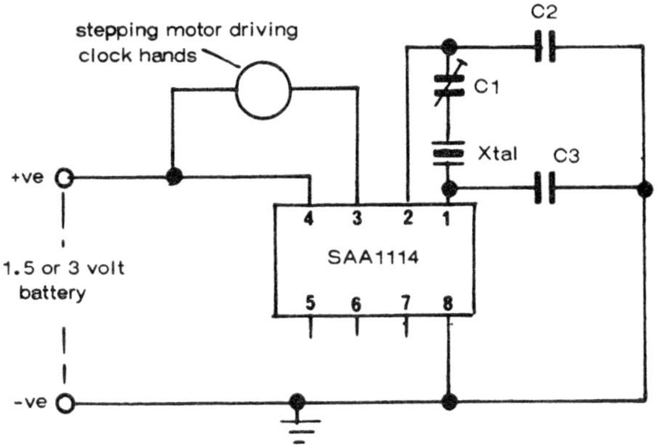

13.16 Crystal controlled clock circuit.
Component values:
C1 – 22 pF trimmer capacitor (type 2222 808 32409)
C2 – 22 pF
C3 – 22 pF
Xtal – 4, 1943 MHz (type no. 4322 143 03111)
IC – SAA1114

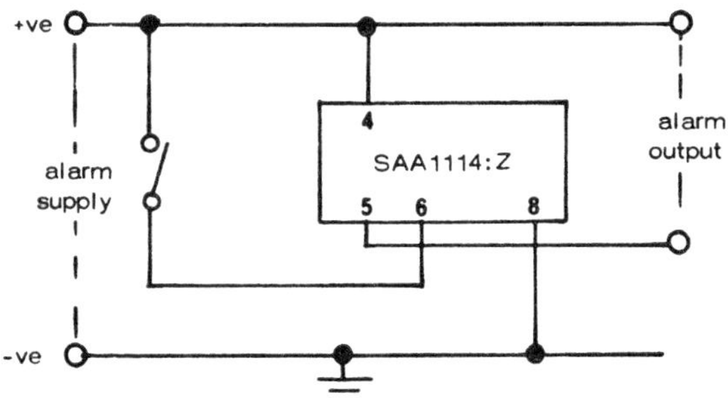

13.17 Additional alarm facility provided in IC SAA1114:Z via pins 5 and 6.

from pins 5 to 6 is a 250 Hz tone signal operating for 4 seconds when the alarm is triggered. External connections for this alarm circuit are shown in Fig. 13.17, the clock motor circuit being as in Fig. 13.16.

INDEX

INDEX